P9-DVP-878

THE VERY FIRST LIGHT

ALSO BY JOHN BOSLOUGH

Stephen Hawking's Universe
Earth: A Photographic Journey
America's National Parks
Masters of Time: Cosmology at the End of Innocence

THE
VERY
FIRST
LIGHT

THE TRUE INSIDE STORY
OF THE SCIENTIFIC JOURNEY
BACK TO THE DAWN OF THE UNIVERSE

John C. Mather and John Boslough

BASIC
BOOKS

A Member of the Perseus Books Group

COLLEGE OF THE SEQUOIAS
LIBRARY

Copyright © 1996 by John Irvan Boslough and John Cromwell Mather.

Published by Basic Books,
A Member of the Perseus Books Group.

All rights reserved. Printed in the United States of America. No part of this book may be reproduced except in the case of brief quotations embodied in critical articles and reviews. For information address Basic Books, 10 East 53rd Street, New York, NY 10022-5999.

Designed by Laura Lindgren

Library of Congress Cataloging-in-Publication Data
Mather, John C.
    The very first light : the true inside story of the scientific journey back to the dawn of the universe / John Mather and John Boslough.—1st ed.
        p.   cm.
    Includes bibliographical references and index.
    ISBN 0–465–01575–1 (cloth)
    ISBN 0–465–01576–x (paper)
        1. Cosmic radiation background.   I. Boslough, John.   II. Title.
QB991.C64M38   1996
523.1'2—dc20   96–10781

98 99 00 01 02 ❖/HC 10 9 8 7 6 5 4 3 2 1

For my mentors, Paul, Pat, Mike, Ray, Jane,
and Nancy, and the COBE team.
—J. C. M.

For James Gibson Boslough, my mentor for this book.
—J. B.

The old men used to study the stars very carefully and in this way could tell when each season began. They would meet in the ceremonial house and argue about the time certain stars would appear, and would often gamble about it. This was a very important matter.

—CAHUILLA (CALIFORNIA) INDIAN STORY

The most beautiful experience we can have is the Mysterious. It is the fundamental emotion which stands at the cradle of true art and true science.

—ALBERT EINSTEIN (*The World as I See It*)

When a man has a certain fantasy, another man may lose his life, or a bridge is built. . . . Everything you do here—all this, everything, was a fantasy to begin with, and fantasy has a proper reality. That is not to be forgotten; fantasy is not nothing. It is, of course, not a tangible object; but it is a fact, nevertheless.

—CARL JUNG, IN AN INTERVIEW WITH
RICHARD EVANS IN *Jung on Elementary Psychology*

CONTENTS

LIST OF ILLUSTRATIONS

ACKNOWLEDGMENTS

M ANY PEOPLE HAVE BEEN exceedingly generous with us with their time, knowledge, or ideas, some of whom put in appearances in the book and some who do not. We cannot begin to thank them all enough, but will try. We appreciate the assistance and support given us by members of the COBE science team, whose calendars often are filled to overflowing: Charles Bennett, Edward Cheng, Eli Dwek, Samuel Gulkis, Michael Janssen, Thomas Kelsall, Phillip Lubin, Stephan Meyer, S. Harvey Moseley Jr., Thomas Murdock, Richard Shafer, George Smoot, Robert Silverberg, and Edward Wright. We appreciate especially the time and thoughts of COBE science team members Nancy Boggess, Michael Hauser, Rainer Weiss, and David Wilkinson, without whose gracious help this book would be far from complete: Together they provided invaluable information and fresh details about the inner workings of the NASA bureaucracy and the early years of the exploration of cosmic background radiation; without their input, many of COBE's significant historical antecedents would never have surfaced here.

Other individuals who work or once worked for NASA as civil servants or contract employees or outside contractors also helped, by sharing their recollections, by reviewing the manuscript, or simply by providing support: Susan Adams, Anthony Fragomeni, David Gilman, Richard Isaacman, Dennis McCarthy, Franklin Martin, John Wolfgang, H. Richard Freeman, Steven Paddack, Martin Donohoe, Robert Maichle, Michael Ryschkewitsch, Michael Roberto, Dennis Evans, Ramesh Sinha, Lawrence Watson, Olof "Bud" Bengtson, Richard Hopkins, Richard Herring, Daniel Payne, and Gerald Godden. We also thank Keay

Davidson, who painstakingly gathered voluminous and invaluable research material; John Brockman and his associates in New York; and Susan Rabiner, senior editor at Basic Books, with whom we participated in an invaluable prebook conference that stayed with us throughout the project. Physicists, astronomers, and others who either helped directly in the preparation of the manuscript or otherwise provided useful ideas or support include: Ralph Alpher, Robert Herman, Arno Penzias, Robert Wilson, P. J. E. Peebles, and the staff of the Niels Bohr Library at the American Institute of Physics in College Park, Maryland.

From John Mather: I extend special thanks to my mentors over the years. My parents encouraged me in a very nondirective way from an early age to follow my scientific inclinations, and my father taught me statistics. Paul Richards at Berkeley taught me experimental physics and designed the apparatus that started us on the road to measuring the cosmic background radiation spectrum, and my fellow graduate student David Woody made our balloon payload work. Patrick Thaddeus at the Goddard Institute for Space Studies encouraged me to develop the COBE idea and pointed me toward the right people. Mike Hauser hired me at Goddard, watched over my entire professional career, and set me some nearly impossible tasks, all the while showing scrupulous attention to detail, compassion, and examples of gentlemanly behavior. Nancy Boggess strongly supported the COBE Project at NASA Headquarters and then came to Goddard to work as deputy project scientist, where she was a wise advisor and a demon for work.

My engineering colleagues who managed and built COBE, the secretaries and business staff who made the NASA system work for us, my fellow scientists who defined the mission, guided the engineering effort, and analyzed the data, my friends at Headquarters who defended and funded the COBE—all were essential parts of the team. Ray Weiss led the Science Team and kept us together despite many challenges. Roger Mattson and Dennis McCarthy led the engineering part of the project, Tony Fragomeni led the spacecraft effort, and Bob Maichle and Mike Roberto led the effort on the FIRAS instrument. Rich Isaacman and Shirley Read led the FIRAS software effort for a long time, Rick Shafer was my deputy principal investigator, and Dale Fixsen told us how to calibrate the FIRAS. George Smoot put COBE on the front page. This book would not exist without them.

This book also clearly would not exist without the writing skill and technical knowledge of my co-author, John Boslough. He took my story of the COBE team and filled it out by his own efforts, interviewing colleagues, reading reports, and putting the human touch on the facts of the matter. He also wrote much of the historical background material directly, building on his previous work with Stephen Hawking and other cosmologists. I learned some of the history for the first time from him. It's been a great pleasure doing this book with him.

My wife, Jane, has been my most important mentor of all. She has a fine instinct for human nature, she is naturally curious about all sorts of things, she is ready to say when things don't make sense, and she makes sure I stay in touch with real life and keep a human perspective on our work. In many ways that is more important than everything else. For her guidance and love I am enormously appreciative.

From John Boslough: I would like to thank those who provided support and/or inspiration or simply helped when necessary. I can't begin to thank them enough: Dr. James Gibson Boslough, Katherine Gibson Boslough, Jill and Jack Karson, Luisa Ilustre, Carlota F. Shea, and little Sophie Elizabeth Boslough, who offered her Daddy relief from the grind of finishing the manuscript in the form of playtime merriment when he needed it most.

Quite obviously, there would have been no COBE project without John Mather, and without COBE this book would not exist. Despite the pressures of his multifaceted job at NASA, John was a most congenial and efficient co-author who amply displayed why he is so good over the long haul. I thoroughly enjoyed working with him on this relatively short-haul (at least in NASA terms) book. I also wish to thank my wife, Susan Raehn Boslough, who occasionally must have felt like a book widow but smiled through it all, while providing just the right amount of love and support along with the perfectly timed offhand killer remark to ensure that this book got finished only a few months late.

It should go without saying, though it will not, that whatever errors have managed to make it through the research, writing, review, and editing processes are entirely the responsibility of the authors, who do not wish to imply that others whose names happen to appear in the text are the offending parties.

PREFACE

This universe is not hostile, or yet is it friendly. It is simply indifferent.

　　　　—JOHN H. HOLMES *The Sensible Man's View of Religion*

O N APRIL 23, 1992, shortly after 8:00 A.M., members of a National Aeronautics and Space Administration team walked into a ballroom at the Ramada Renaissance Hotel-Techworld in Washington, D.C. There they first revealed to the world findings made by a remarkable satellite called the Cosmic Background Explorer, or COBE. The announcement—that the satellite's orbiting instruments had detected tiny variations in the cosmic background radiation—electrified the global cosmological community, and drew headlines around the world. Many astrophysicists believed that the discovery confirmed the reigning cosmological model, the Big Bang theory, beyond the shadow of a doubt, perhaps solving the mystery of the universe's origin once and for all. Stephen Hawking, the Cambridge University theorist and best-selling author of *A Brief History of Time,* was the most effusive of all: "It is the discovery of the century, if not of all time."

High praise, indeed. The reason for Hawking's exuberance was that the COBE discovery had gone right to the heart of cosmology, the science that seeks to explain the very origin and structure of the universe. Did the universe begin at a specific point in time, or has it always existed? For thousands of years most scientists regarded this question as one beyond their concern, lying within the metaphysical realms of philosophers or theologians. Not until the middle of this century did physicists and astronomers begin acquiring theories powerful enough and experimental equipment sensitive enough to begin addressing the problem. COBE had shown emphatically that the universe is not static, but has changed remarkably over time.

In January 1993, John Mather, COBE's project scientist, announced more startling news sent back to Earth by the COBE satellite. The latest data showed that at least 99.97 percent of the early radiant energy of the universe was released within the first year after the Big Bang itself. The new COBE discovery—in the form of hundreds of millions of measurements—severely limited the size of "aftershocks" that might have followed the Big Bang, making it unlikely that there were many "little bangs," as some theories have suggested. "The closer we examine the Big Bang the simpler the picture gets," John Mather explained in a press release accompanying the announcement. Taken by an instrument specially designed to receive energy directly from the Big Bang itself, the data provided the closest scientific look ever at the moment of creation. As the years pass, COBE's findings will be studied and analyzed intensely. Our understanding of how the universe was created and how it evolved surely will be enhanced; at the same time, new constraints might have to be placed on existing theories purporting to explain the structure of today's universe.

Although there had been earlier important antecedents, COBE's story really began in 1964. That year Robert W. Wilson and Arno A. Penzias of Bell Laboratories, while trying to eliminate microwave noise from an antenna designed to pick up satellite signals, discovered a phenomenon known as the cosmic background radiation. The radiation suffused the sky in every direction at microwave frequencies, and came to be widely regarded as the afterglow of the Big Bang. In the aftermath of this discovery, cosmology became an empirical science. During the 1970s and 1980s, more sophisticated equipment refined the measurements of the cosmic background at different frequencies. But a major puzzle remained unsolved throughout these years: The microwave background seemed too smooth. It lacked the slight variations in temperature and, by implication, in density that appeared needed to seed the gravitational clumping we see in the later universe. Without such seeding, cosmologists believed, there would have been insufficient time to produce the galaxies and immense clusters of galaxies being observed in the universe today.

Enter COBE—or at least the first glimmer of an idea for it. In the early 1970s, while completing graduate work in physics at the University of California at Berkeley, John Mather participated in a difficult experiment that helped establish the spectrum of the cosmic background radiation. In 1974 he responded to a call from NASA for proposals of cosmology experiments that could be carried into orbit by the space shuttle, gathered a team of seven scientists, and sketched out a rough diagram for a satellite. This early design featured all three of the experiments that eventually were carried into orbit by the COBE satellite. Over time, NASA added researchers to the original team, including George Smoot, then at Berkeley, and Samuel Gulkis at the Jet Propulsion Laboratory in Pasadena, while others dropped out. But the three experiments—each consisting of a specialized instrument— remained the heart of COBE.

The three instruments exceeded all expectations. Few people, including scientists, may realize, however, that by the time COBE scientists began making their series of announcements, many team members had been working full-time on the project for nearly two decades, some devoting their entire working lives to it and staking their careers on its success or failure. This book is not only an insider's account of the COBE mission, along with its many technological and bureaucratic tribulations during the 1980s (including the horrifying explosion of the *Challenger* space shuttle). It is also the story of the dramatic and serendipitous discovery of the cosmic background radiation itself, told in a new light, and of how that discovery led to the modern era of cosmological investigation.

Since science is not a cut-and-dried endeavor, but one deeply dependent on the hopes and dreams of the people who pursue it, in this book we tell the story of some of those people: how and why they came to practice science, how scientific credit—the driving force behind scientific endeavor—is dispensed, sometimes haphazardly, and how people react when that credit is given or withheld. Careers were established while some endured bitter disappointment as the result of a still lingering controversy over who deserves scientific credit for the discovery of the cosmic background radiation. Often, we will see, the path to the truth is convoluted, with no clear answer emerging.

Nearly a quarter of a century after the discovery of the cosmic background radiation, the COBE team found itself embroiled in a new controversy over scientific credit. Most COBE scientists believed that over the project's lifetime they were part of a genuinely amiable, cohesive, well-oiled team. Yet as the time came for the team to announce to the world the stunning data its spacecraft had sent back to Earth and it became apparent how eagerly that news would be received, one team member broke ranks and stepped into the limelight alone. How that unfortunate series of events played out is an important part of the COBE story.

It is worth noting that a few cosmologists, certainly a minority, feel that the COBE discoveries had little to do with the origin of the universe. One that should be mentioned because of his diametrically opposite view and because of his stature in the scientific community is the late Swedish Nobel laureate Hannes Alfvén. For the last years of his life Alfvén led a vocal minority maintaining that the universe has always existed—no Big Bang required—and that it has been shaped more by electromagnetic currents than by gravity. In Alfvén's scenario, the microwave background is caused by the continuous emission of electrons circling in magnetic fields jetting out from billions of galaxies. The microwave background is thus a sort of radiation fog whispering through the cosmos, and the temperature irregularities discovered by COBE are only to be expected. A few other physicists such as J. Anthony Tyson at AT&T's Bell Laboratories think still another kind of mechanism—maybe black holes—could account for cosmic structure. Lawrence M. Krauss and Martin White of Yale have argued that the variations in the cosmic microwave background could be just an expected consequence of Albert Einstein's theory of relativity, which is that the distortions could be caused by gravitational waves. Indeed, the COBE results provide the best answer to such speculations.

Still, most cosmologists believe that COBE has improved our knowledge of the universe many times over, and that its satellite successors, now in planning, will continue to do so. Indeed, we have high hopes that COBE data still being analyzed as well as data from future satellites will startle us anew with more dramatic information from the dawn of time. If so, that would renew the debate about the most fundamental question of all, "Where do we come from?" After all, we are all cosmologists staring out at eternity from a small corner of the universe with the hope that at last we are beginning to understand it.

• • •

This book grew out of a pleasant afternoon discussion in 1992 between the co-authors at NASA's Goddard Space Flight Center in Greenbelt, Maryland. It by no means tells the full COBE story, which at the project's peak involved more than 1,500 hardworking individuals and is so rich in untold, and undoubtedly never to be told, incidents of individual effort at unknown crucial junctures that no single volume could include it all. The authors apologize to those many fine engineers, lab technicians, support staff members, NASA officials, and others whose names do not appear in the text. COBE would never have flown its journey of discovery out into the universe without your enthusiasm and excellent work.\

We decided to tell the story of COBE from the first-person perspective of John Mather, who initiated the project with a proposal to NASA in 1974 and two years later became COBE's study scientist and then-project scientist, or scientific leader; he has spent his entire working life on the project and was directly involved with it from start to finish. For those parts of the story relating to either COBE or the history of the cosmic background radiation in which John Mather was not present, we have relied on third-person accounts.

PART I

A Day Without Yesterday

Better Is the Enemy

We will end up where we are going if we do not change direction.

— CHINESE PROVERB, 6TH CENTURY B.C.

VANDENBERG AIR FORCE BASE perches strategically above the Pacific Ocean like a giant condor guarding its aerie, its 153 square miles of desert and mountains not particularly inviting. An old army camp dating from before World War II, it had once been called Camp Cook but had been renamed in honor of an Air Force chief of staff. Signs along the base's 35 miles of coastline warn visitors away from unpredictable riptides and crumbling cliffs. Elsewhere you are advised not to pick up or take home as a souvenir any "suspicious object," which could be an unexploded bomb. With this sense of past and present danger all around and with its barren landscape, Vandenberg gives the impression that it is a kind of national park or wilderness refuge in reverse—which is more or less what it was when it was used as a prisoner-of-war camp during the war.

Vandenberg serves today as the West Coast site for launching military satellites into the pole-encircling orbits that provide the best vantage for observing Russia. Since the early 1960s approximately 1,700 rockets have been fired into the sky from the base's dozen or so widely separated launch pads, a few within half a mile or so of the surf pounding the coast. Almost every rocket from the base is launched south over this coast. If one blows up, it plops harmlessly—

although not inexpensively—into the Pacific. Vandenberg also launches scientific satellites requiring a polar orbit. Along with numerous colleagues from around the country, I was deeply involved in the design and construction of such a satellite for fifteen years. The moment of truth had arrived at last: Our satellite was scheduled to be launched early Saturday morning, November 18, 1989.

I went out to Vandenberg a week early to help review last-minute preparations. The countryside as I drove up the final few miles from Santa Barbara was desolate but lovely: deeply eroded, silvery hillsides bristling with patches of scrub brush and live oak. Lompoc is a little community at Vandenberg's south gate. With friendly shops and restaurants along an old-fashioned Main Street and tree-lined residential avenues, the town of 38,000 exists mainly to serve the rocket base. I was delighted by what I saw. And smelled. Lompoc is the center of a flower-farming district, and there were huge fields of flowers everywhere even in late autumn. After checking into the Porto Finale Inn, I drove over to the base.

Along with its extensive launch facilities, Vandenberg contains an off-limits intercontinental ballistics missile range that was on constant alert during the cold war years. I drove to launch site SLC-2W, an immense building that looked like a hangar at least ten stories tall in the midst of a barren desert landscape with hills and mountains in the distance. Members of our team who had been there for several months offered me a cook's tour. Inside was an unforgettable sight. Surrounded by a web of steel girders was a Delta Rocket type 5920, 116 feet tall and marvelously beautiful to my eyes despite its many welded patches. Others on the team were less certain of its beauty.

Although I had worked as a scientist for the National Aeronautics and Space Administration for nearly fifteen years, this was my first launch. My previous experience was limited to the gleaming white rockets we've all seen on television when the space shuttles or the Apollo expeditions to the Moon were fired into the sky. This rocket had been put together from spare parts from other Deltas. Some of the pieces had been left outside for years and pigeons had defecated on them, the caustic material creating holes in the aluminum. These had been patched here and there with welded sheets of aluminum, making the Delta look a little less than pristine. Yet the sight was not especially worrisome. Engineers from McDonnell Douglas, the company that managed most of the facilities at Vandenberg, had assembled the rocket and would launch it under NASA supervision. I had the utmost faith in them, knowing they were seasoned professionals and realizing that the success or failure of the project was now in their hands and well out of mine. They had already launched nearly 200 Deltas with only 4 failures.

A few days later, one of the engineers escorted me and a few scientific colleagues—Mike Hauser, Nancy Boggess, Ralph Alpher, and Robert Herman—up the gantry elevator. The overriding sense was of danger—not from the height but from the fact that the rocket would be packed full of a frightening amount of explosives. There would be more than 200 tons of liquid fuel for the main engine at launch and a powerful solid fuel in the booster rockets. It was impossible not to imagine what it would be like to be an astronaut sitting atop such a stack of explosives waiting to be launched into space. The thought was irrationally terrifying, although it was obvious that this was a matter of routine by now, sixty-three years after Robert Goddard had first tried to launch a small rocket in a field.

The elevator stopped. The sum total of what many of us had invested our entire working lives in was encased within a protective aluminum shroud on top of the Delta. The Cosmic Background Explorer had been shipped the month before from Andrews Air Force Base near Washington, D.C. Many of us who had worked on it believed it had the potential to become one of the most scientifically fruitful unmanned missions of the space age. But now, during its final preparations for launch, the COBE (rhymes with "Moby") looked surprisingly mundane. The size of a small truck, it weighed two and a half tons. Its three solar panels, which would be opened in orbit, were folded neatly along its sides, as was the thermal/RF shield. It looked as though it were getting ready to go to work.

Once in orbit, the COBE's job would be to scan the sky and send back 4,096 bits of data per second that we believed would advance—*greatly* advance, many of us hoped—one of the most exciting scientific endeavors of the twentieth century: the quest to understand how the universe began and how it has evolved since. That humans could even contemplate supplying answers to such questions filled me with awe. And yet looking at COBE bolted onto the Delta rocket, I thought mostly of the 5 million people-hours that had gone into its construction, the hundreds of millions of dollars expended on it, and the 1,600 people at NASA's Goddard Space Flight Center and around the country who had worked on it. Although Deltas had failed, the rocket's success rate was 98 percent—one of the best launch vehicles in the world, NASA rocket experts had assured members of our scientific team. Still, unlike the rockets used for manned missions to space, it was a mostly nonredundant vehicle. It had only one of everything: one on-board computer, one tracking system, one electronic system, and so on. Would our Delta hold together under the tremendous stress of the launch? In the previous few years there had been several major launch explosions, including a Titan, a Delta like ours, and even the *Challenger*. I wondered what the fates had in store for us.

Nancy Boggess, an astronomer from Goddard and a fellow member of the science team that had conceived COBE and directed its development, had also arrived early at Vandenberg. For the next few days we attended review sessions of each of the many complicated systems of the Delta and the Cosmic Background Explorer with the engineers and technicians who would launch it: the main engines and the nine solid-fuel booster rockets that would propel COBE into orbit, the fuel systems, the telemetry, and the tracking equipment. At the conclusion of the reviews we met with the engineers. Although it was impossible to prove that everything was 100 percent ready, I recalled a phrase engineers liked to cite: "Better is the enemy of good." It had impressed me enough to look it up; it was from Voltaire.* I agreed with the engineers and technicians that everything looked "good enough" to go. There really wasn't any way to worry creatively about something so enormous. We had done everything we could.

A day or two before the launch, other members of our science team began arriving. George Smoot, the principal investigator for one of the three experiments aboard COBE, drove down from Berkeley. Sam Gulkis and Mike Janssen came up from the Jet Propulsion Laboratory in Pasadena, Ned Wright from UCLA, Rainer Weiss from MIT. There was Dave Wilkinson from Princeton, along with Mike Hauser of Goddard, another principal investigator, all the other science team members, and many other colleagues from Goddard. Unfortunately,

---

* "Le mieux est l'ennemi du bien," Voltaire wrote in *Dictionaire Philosophique, Art Dramatique,* 1764: "The better is the enemy of the good."

many of the major players couldn't attend the launch because they remained back at Goddard ready to begin operating the spacecraft right after launch. The media began appearing, along with hundreds of invited guests. My wife, Jane, and her mother, Anne, drove up from Santa Barbara.

We held a press conference at the hotel the afternoon before the launch. More than 100 NASA staffers and colleagues, newspaper and television reporters, and guests attended. McDonnell Douglas official Don Tutweiler went over details of the launch, reassuring everybody that it would go off without a hitch. As project scientist, I briefly explained the purpose of the mission. Two physicists in their sixties, Ralph Alpher and Robert Herman, also spoke. Forty-one years earlier, working with George Gamow, they had made a scientific prediction about the birth of the universe that helped lead to COBE's mission into space. This prediction, which had been accurate, as later events were to show, was that a phenomenon known as the cosmic background radiation—a type of microwave energy—fills the entire universe. We were going to measure it extraordinarily well, if everything went off as planned.

Most astronomers and cosmologists now regard this microwave radiation suffusing the sky in every direction as the afterglow of the Big Bang. According to this theoretical model, the universe evolved billions of years ago from a state so hot and dense that only radiation and elementary particles could have existed. Although the Big Bang is loosely described as the beginning of the universe, or even as a creation event, that has never been a satisfying picture for me. I would like to know what happened before that. Physics has no words to describe creation from nothing, but describes only the transformation of one substance into another. Impossible as it may seem, there is real hope of understanding the Big Bang in terms of even more exotic but probably yet unknown laws of physics. To do our part, we had designed the COBE to carry three sophisticated scientific instruments to make sensitive measurements of the radiation from the early universe that still remains. We hoped the data sent back from the COBE would help answer the most fundamental questions: What were the conditions like when the universe "began"? How did it become the universe we live in today?

Alpher and Herman were delighted to have been invited. They had believed for years that they had not been given adequate credit for their prediction. In speaking at the press conference, they felt their work at last was being acknowledged in a public forum. Alpher gave a brief synopsis in lay terms of the development of their work back in the 1940s, then explained how scientific ideas about the origin of the universe had evolved since then. He praised the COBE as a "marvelous instrument that we hope will provide the most accurate information we have ever taken about the origin of the cosmos."

That night several of us drove over to the small town of Santa Maria, where we attended a dinner meeting of science teachers. Larry Caroff of NASA Headquarters in Washington, D.C., spoke, saying he believed the observations taken by the satellite would "give us an unbelievable view of the universe. We'll see things nobody's imagined. This could shake up the world of science." Don Tutweiler, the project manager for the Delta at McDonnell Douglas, gave a wonderful talk about the launch itself. He explained how the engineers would have to "gently back the rocket off from the COBE at 18,000 miles per hour because these scientists over here are worried that we might damage their satellite." I spoke at some length to explain the scientific importance of what we were attempting, and to help the teachers explain the concepts to their students.

The launch was scheduled for before dawn the next morning. We were to awaken at three o'clock, and tried to get to bed early without success. After what was probably for most of us an almost sleepless night, we were picked up by buses that took us out in the predawn darkness to observing sites scattered around the launch pad. Nearly two thousand invited guests, the largest turnout in Vandenberg history, would witness the spectacle. The viewing area for the press—where about fifty of us were taken—was a hillside a mile or two from the launchpad.

At that distance the Delta, powerfully illuminated in the early morning darkness, was only about the height of a fingernail at arm's length. Coffee and doughnuts were served from an old trailer as we huddled under a few lightbulbs strung from short wooden poles, trying to stay warm in the early morning cold. In the group were Alpher and Herman and my former doctoral thesis adviser at the University of California at Berkeley, Paul Richards.

Shortly before, launch engineers had replaced the Delta's faltering on-board computer with a new one. We had no Plan B in the event the rocket exploded. There had been some talk from NASA Headquarters of building another COBE if this occurred, but we had not formalized such an idea: I would have considered it bad luck and a waste of effort anyway. In any event, I had little time for such contemplation. I was too busy talking with newspaper and television reporters, spending most of my time with David Oyster, a producer for a PBS series called "The Astronomers."

Although we were scarcely aware of it, the countdown had begun. Weather balloons already had been sent aloft to test high-altitude winds. They indicated that upper-air wind shear was too great, and the countdown was halted. More balloons were launched. We waited and waited. The countdown resumed, and there was relief all around. The launch window, the period during which the rocket could be fired in order for the COBE to be placed in the proper polar orbit, was only about thirty minutes long.

With little warning, a great flash of light lit up the sky. The 440,000 pounds of rocket began lifting off at what seemed to me an agonizingly slow rate. We could hear nothing. Sound waves never seemed to travel so slowly. The silence of the Delta's ascent into the black sky was eerie, almost frightening. Sound waves from the blast had been suppressed by thousands of gallons of water pumped rapidly through a concrete channel directly beneath the rocket. This standard launch procedure was designed to prevent powerful vibrations from traveling upward and damaging the rocket as it rose.

By the time we heard—and felt—the blast from the launch, the rocket was traveling nearly as fast as the sound waves now rattling us. Within a minute and a half, the rocket was more than 15 miles up, and traveling nearly a mile a second. Its white contrails, glistening brightly in the light of the rising Sun while we watched from the darkness below, curled majestically around a quarter Moon. There was little doubt that the first part of the launch had been perfect. As we watched in silence, tiny dots appeared to break away from the rocket. Six of the solid fuel-booster rockets had been ejected. Thirty seconds later the rocket was more than 25 miles high and barely visible. It expelled its remaining three solid fuel boosters, which splashed into the Pacific some 230 miles to the south.

Four minutes after launch, as the Delta's first stage plummeted toward the sea, the second stage fired. A few seconds later the fairings protecting the Cosmic Background Explorer were ejected, and it was exposed for the first time to the cold vacuum and fierce sunshine of empty space some 70 miles high. Five minutes later the rocket's second-stage engine was automatically turned off by the on-board computer and COBE cruised silently in orbit 105 miles above the surface of the planet it had departed from only a little more than ten minutes before.

Many of us were too exhausted from the sleepless night before, and no doubt from nervous energy too, to feel much elation. Still giving interviews to reporters, I made my way back to the hotel for the postlaunch party, complete with champagne, that was being hosted by NASA's public affairs office. The party was almost over by the time I got there. Only then, a little more than an hour after launch, did we learn from a jubilant Dennis McCarthy, COBE's deputy project manager, that the Delta rocket's second-stage engine had successfully reignited, propelling COBE into its final orbit: a nearly perfect north-south circle around the earth 559 miles high. Our COBE took its first action as helium valves opened to start cooling two of the instruments.

COBE was now a satellite. There was nothing more for members of the science team to do in California. I left the party and went to collect family members, and we went out for pancakes. Jane and I drove back to her mother's home in Santa Barbara, stopping at the ancient La Purissima mission to have a little privacy and think about the meaning of life before technology. Then we drove to the airport for the trip back to Washington.

Everybody was exhausted, but my colleagues and I were all eager to get back to Goddard. We expected to learn there whether the COBE detectors we had spent so much money and effort placing into orbit really would be able to tell us anything new about the infancy of the universe.

✳ 2 ✳

Scooped

Before we can discuss the basic problem of the origin of our universe,
we must ask ourselves whether such a discussion is necessary.

—GEORGE GAMOW, *The Creation of the Universe*

PEOPLE ON OUR PLANET have always been explorers of the unknown. For more than 99 percent of our history, this has meant journeying from one corner of our planet to another, foraging for food or hunting it down. When the food ran out or the river ran dry, we moved on. Our journey down through the ages has been a story of great migrations: of Africans to India and the great land mass of Asia, of Indians to Africa, of Asians to North America, of North Americans to South America, of Mediterranean people to Europe. Only recently in our long history— the last ten thousand years or so—did we discover that we could grow our own food. At last the wandering tribes settled down, becoming farmers instead of gatherers and hunters, domesticators of animals and the plants used for food, sedentary, no longer adventurers in constant search of sustenance.

Perhaps as we farmed we became bored, restless, edgy for the open road, the open prairie, the open sea. No longer in search of food, we started traveling to new lands for the sake of exploration alone, climbing a hill or a mountain range or canoeing across a river for the sheer joy of seeing what lay on the other side. Later adventurers undertook dangerous journeys to distant lands and across the seas for the material riches their exploration could bring. Greek soldiers led by

Alexander of Macedonia waged a journey of conquest overland into Asia; Chinese explorers in large junks may have crossed the Pacific and landed on the shores of North America; people from the Indonesian archipelago sailed northward in oceangoing outrigger canoes, eventually landing and settling in Hawaii; and, centuries later, Europeans in sailing vessels circumnavigated the globe.

Needing guidance, the explorers relied on the ancient sky, the single visible constant in the life of every person who ever lived on Earth. No culture ever ignored it. For the Anasazi of the Southwest, the sky was the "backbone of the night," and they built stone astrolabes to chart its movements; for ancient Scandinavian sailors the sky, in its regular but mysterious comings and goings, was "the path of the ghost." Astronomy was the first science in the ancient civilizations of Greece, Sumeria, the Nile, China, and Mesoamerica, generally accompanied by astrology, a pseudoscience that exploited the mind's eagerness to link unrelated events.

A remarkable piece of evidence for our ancient fascination with the sky stands in ruins in northern Kenya at a place called Namoratung'a: a consciously designed array of nineteen stones that was one of the first observatories in Africa. It may have been built by astronomer priests who came from eastern Asia about three thousand years ago in a great migration of the Galla people to northern Africa.[1] The observatory was probably constructed to predict the conjunction of the new Moon with several significant patterns the builders saw in the sky. The carefully placed stones marked the horizon position of the rising Moon and the constellations the astronomer priests used to formulate their calendars. The observatory at Namoratung'a worked well for only about a century, owing to the precession of the equinoxes. An astronomical phenomenon caused by the gravitational attraction of the Sun and the Moon on the earth's equatorial bulge, the precession makes the stars appear to rotate westward in a grand cycle taking 25,800 years (the same time span, incidentally, ancient Chinese astronomers calculated as the great circle of life, after which history would start all over again).

Once the predictions of the stones were no longer fulfilled, the Namoratung'a skywatchers abandoned their observatory. This would happen throughout history: As astronomers encountered unexpected patterns in the night sky, they would have to give up—always resisting the change in thinking—their cherished models. In the fourth century

B.C., Aristotle created a comprehensive model of the cosmos based on the natural assumption that the earth lay at the center of the universe, the sky revolving around it once each day. Ptolemy, the Greco-Egyptian astronomer in Alexandria during the second century A.D., refined the Aristotelian model of the universe and codified it in the *Almagest*. Astronomers relied on it for more than a millennium to calculate the positions of objects in the sky as they were carried about the earth on individual rotating spheres. Commenting on its complexity when it was explained to him, Alfonso X (The Wise), king of Castile and Leon (1252–1284), reportedly said, "If the Lord Almighty had consulted me before embarking upon Creation, I should have recommended something simpler."

That the Sun might not circle the earth simply must have defied logic. In 1508 the Polish astronomer Copernicus, realizing the Ptolemaic patterns in the sky had shifted, dared write the words: "What appears to us as the motions of the Sun arise not from its motion, but from the motion of the Earth." Such an idea was dangerous, alien to the religious and philosophical thinking of the day in which humankind occupied the physical and metaphysical center of the universe. "The fool will turn the whole science of astronomy upside down," wrote Martin Luther.[2] That is exactly what Copernicus accomplished when, after years of doubt, he finally dared publish *De Revolutionibus Orbium Coelestium,* his formal declaration of a heliocentric solar system. The publication occurred the year Copernicus died, the year Galileo Galilei was born, and the year some historians mark as the beginning of modern science: 1564.

Late in the following century, Isaac Newton initiated the true scientific revolution when he explained how the force of gravitation alone was responsible for maintaining the planets, including Earth, in their orbits around the Sun. The next major overhaul in our astronomical view of the universe did not take place for three more centuries, at the onset of the twentieth century. This was to be an era unlike any other for the exploration of the sky, an era during which we realized for the first time how high the sky really was and how small our own corner of the universe.

A major milepost in this exploration occurred almost exactly twenty-five years before launch day at Vandenberg. The COBE spacecraft traces its ancestry directly back to this event. At the time astronomers had a single overriding question: Has the universe always been, with an infinite age, or did it come into being at some point in time with a probable age of ten to twenty billion years? In 1965 there were good arguments on each side, but neither side had proof. That year a meeting of five men at Crawford Hill, New Jersey, helped resolve the question in the minds of most scientists (although a few determined holdouts still remain today). The meeting also set the tone for the scientific examination of the origin and structure of the universe for the rest of the twentieth century.

The meeting was the culmination of an unlikely sequence of coincidences, and the site was unlikely as well for what was an extraordinary moment in the history of science. Crawford Hill was a research facility of Bell Telephone Labs, Inc., located in a wooded area in the central part of the state near Holmdel. Bell Labs was an enormous facility with 25,000 employees and a budget, privately funded by AT&T, as great as 15 percent of that of the entire National Science Foundation. In those days Bell was a monopoly and could afford to undertake research with long-term or unpredictable payoffs. Its laboratory was the world's leading industrial research facility.

Fostered by leaders with vision, Bell Labs had an open, university-like freedom that encouraged scientists working there to pursue interesting, if not necessarily commercially fruitful, topics. Had Bell Labs, General Electric's laboratories, and others like them been subjected fully to the Darwinian competition and often creative destruction of modern capitalism, the world would be a different place today. The transistor, invented at Bell Labs, might not have appeared for decades; the computer revolution sweeping the world might still be in the future; and Americans could be importing technology instead of exporting it. A few Bell scientists worked as astronomers or theorists in relatively arcane areas like particle or materials physics. But the main emphasis was on good, basic research in more immediate technology such as radar or the new telecommunications computers then being developed.

Arno Allan Penzias arranged the meeting that was to change the way we perceive the universe and lead to the discoveries nearly thirty years later that are the subject of this book. Penzias had been hired by Rudolf Kompfner, director of Bell's Radio Research Lab, in 1961, the year before Penzias was to receive his doctorate in physics from Columbia University. The son of Jewish émigrés from Germany, Penzias was a lean man of great energy and roving intelligence. He was born in Munich on April 26, 1933—the day the Gestapo was created, although, as he likes to observe, he's lasted longer. He graduated from City College in New York City in 1954. Then he struggled through the physics graduate program at Columbia, which claimed as professors such lights as Nobel laureates I. I. Rabi and Polykarp Kusch and other European immigrant scientists who had fled the Nazi and Communist regimes. In those days Columbia was notorious for being tough on its physics graduate students, flunking them out right and left with little apparent discretion.

Penzias's first assignment at Bell was to help solve a problem that had been baffling laboratory scientists: how to point an antenna accurately at Telstar, the first working communications satellite. The antenna was big and heavy, and bent under gravity, wind, sunlight, and temperature variations. This made it virtually impossible to keep it focused on a distant satellite. Penzias figured out almost at once how to solve the problem. He placed a small optical guider telescope inside the larger one and aimed it at a specified star. By comparing the star's known position with its position as reported by the small telescope, he could determine how far off the large receiver was from the communications satellite and make the appropriate correction.

After successfully implementing this new technology, Penzias decided it was time to get back to radio astronomy, the subject of his Columbia dissertation. This turned out to be a lucky decision because at about the same time, Bell Labs abandoned the communications-satellite business to the Communications Satellite Corporation, or COMSAT, which was then being formed. As its name suggests, radio astronomy is the detection and study of radio waves from space. Radio astronomers analyze the waves to determine their source and energy spectrum. Like any wave at the seashore or in the electromagnetic spectrum, a radio wave can be described by a wavelength that would typically be the distance between two points on the wave where its oscillation is the greatest. Radio wavelengths of a few millimeters to about 10 meters (or from about ⅛ inch to about 30 feet) are able to pass through the earth's atmosphere without being reflected back into space by the ionosphere or absorbed by atmospheric molecules.

A visible lightwave is so minuscule by comparison—only about half a millionth of a meter long—that the universe of the radio astronomer is vastly different from that of the optical astronomer. A few radio sources like our Sun, its radio signal powerful only because of its nearness, and the interstellar cloud of seething hydrogen known as the Orion Nebula are, of course, highly visible. But until the advent of radio astronomy, a huge part of the universe, including startling phenomena such as supernova remnants, pulsars, and quasars, was invisible to astronomers.

In the early 1930s AT&T was just getting into the radio telephone business for overseas service. A young radio engineer named Karl G. Jansky was assigned the task of studying the nature of radio static in order to eliminate it. Jansky needed to record and measure it. For the task he decided to use a new kind of rotating antenna that could pinpoint the direction from which a radio signal emanated. After several years he had isolated two kinds of apparently permanent static. One was thunderstorms, the other an unidentifiable continuous buzz that varied regularly in intensity throughout the day.

Jansky's attempts to identify the second source failed repeatedly. An astronomy graduate student from Princeton University named A. M. Skellet happened to be working temporarily at Bell Labs. Skellet heard of Jansky's problem and went to him with a possible explanation: The regularity of the strange static would be expected if its source were stars that were fixed in the sky in relationship to the earth's rotation. Jansky took new measurements that indicated Skellet was precisely right. The persistent hiss indeed came from the stars—in fact, from the center of our own galaxy, the Milky Way.

Jansky wrote three papers explaining what he had found. They were published in a technical journal called *Proceedings of the Institute of Radio Engineers* instead of in an astronomical periodical, and hence were largely ignored by astronomers.* Bell officials were convinced—correctly—that nothing could be done about the steady stream of static from space and pulled Jansky off the research. He apparently all but forgot about it, along with just about everybody else. In fact, there was no interest in it until eight years later, in the months before the United States entered World War II. That year, 1941, a backyard astronomer

---

*"This is the longest distance anyone ever went looking for trouble," *The New Yorker* magazine reported after learning about the static from the Milky Way. Otherwise, interest in the popular media was negligible.

named Grote Reber, who had read Jansky's papers and become intrigued, built a thirty-foot bowl antenna in the yard of his home in a Chicago suburb. First he confirmed Jansky's results. Then he made a detailed map of the sources of the radio waves that were coming from the direction of the Milky Way.*

As the war progressed, J. S. Hey and fellow scientists at the Jodrell Bank Experimental Station in Cheshire, England, were disconcerted by a continuous noise in their radar equipment. They assumed that the Germans were transmitting some kind of signal to jam British radar. They discovered after months of investigation that radio waves from the Sun and the same persistent Milky Way signal that Jansky had identified years before were actually doing the dirty work. Out of wartime necessity, Hey and his colleagues did some research and resuscitated Jansky's and Reber's work. Bell Labs also had undertaken radar research during the war, and scientists there also dug up Jansky's old papers. Thus radio astronomy, conceived a decade earlier and all but unknown to the global astronomy community, was finally born.

Bell Labs officials eventually decided radio astronomy would bear little fruit. By the time Penzias decided to begin working again in the field, Bell had all but abandoned it to other research facilities and had only one assigned position for a radio astronomer. Penzias took it, and was given a unique receiving antenna to work with. The antenna sat atop the kind of low rise that is considered a hill in central New Jersey. It had been built by Arthur B. Crawford, head of Bell's Radio Research Department, in 1960. About 20 feet long and resembling a huge, hollow horn with the end cut off by a diagonal reflector, the antenna was constructed specifically for receiving the relatively weak signals bounced off the Echo 1 satellite, a prototype for the first communications satellite. This was a huge metallic spherical balloon that the National Aeronautics and Space Administration had launched into orbit on August 12, 1959, aboard its first Delta rocket. This was almost exactly thirty years before the successful launch of the Cosmic Background Explorer aboard Delta Flight No. 189, which was a powerful two-stage rocket rather than the relatively small Delta that carried Echo 1 into orbit.

After finishing its work with Echo, the big horn antenna was trained on Telstar, the first true communications satellite, which Bell

---

*Jansky's name is now a unit of measurement of intensity used by astronomers.

Labs had designed and NASA had launched in 1962. After showing that satellites could be used effectively for transmitting telephone and television signals overseas, the antenna had effectively been put on hold while new ground stations were built in the United States and Europe. When Penzias switched back into radio astronomy in 1963, he persuaded Bell officials to let him convert the antenna from a satellite communications receiver into a radio telescope. Kompfner, Penzias's division chief, apparently had the foresight to recognize that work in radio astronomy might eventually benefit satellite communications.

In the spring of 1963 Penzias was joined in his new position by Robert W. Wilson: Since there was only one job for a radio astronomer, they would split it and take on other individual responsibilities the rest of the time. Wilson was a mild, laconic Texan. Born in Houston on January 10, 1936, he was the son of a chemical engineer who worked in the oil fields. Wilson had been educated in the Houston public schools, then gone on to Rice University intending to become an electrical engineer. But he did not like Rice's engineering department. About the same time, he had begun displaying a substantial enough talent for math and physics to gain him acceptance into the graduate physics programs at the Massachusetts Institute of Technology as well as the California Institute of Technology.

He picked Caltech. There he took a course in quantum mechanics taught by Murray Gell-Mann, who soon would win a Nobel Prize before the age of forty for his ingenious theory of the quark—a subatomic quantum particle that physicists still believe to be among the smallest indivisible units of energy and matter. At Caltech, Wilson enrolled in a course in cosmology, which at the time was a relatively new branch of astronomy specializing in the origin and evolution of the universe. The course was taught by Fred Hoyle, a British astrophysicist already famous as the leading proponent of a cosmological theory that held that the universe had no specific beginning point in time. Instead, he and his supporters argued, the universe was evolving over infinite time in what Hoyle called a "steady state."

The opposite view, about equally in vogue among astrophysicists in the late 1950s, was that the universe had "begun" at a finite moment billions of years ago in a hot, massive expansion from almost nothing. Hoyle had an unusual gift for nomenclature. Several years earlier he had jokingly—and, as time was to prove, ironically—coined the name that would stick for this opposing theory from a British slang expression for orgasm: the Big Bang.

SOMETHING WRONG

At Caltech, Wilson also encountered an Australian radio astronomer named John Bolton and became interested in the field, finally working on a new radio map of the Milky Way for his dissertation. By the time Wilson joined Penzias at Bell Labs in the spring of 1963, Penzias had been working on the big horn antenna for about a year. The first thing they had to agree on was what kind of science they were going to use it for. But first they needed to prepare their highly sensitive instrument. This meant getting rid of excess static, or what is known in the radio business as "noise."

The summer after Wilson arrived, he and Penzias added to their antenna a device known as a cold load. This was a refinement designed to provide a continuous amount of precisely measured low-level noise with which they would be able to calibrate their instrument. In theory, the cold load could be seen as a substitute for the antenna and the sky, in order to allow Penzias and Wilson to calculate the source of any extra noise coming out of the receiver. One of the strongest radio sources then known to exist in the sky was a supernova remnant called Cassiopeia A (in the constellation Cassiopeia). Penzias and Wilson decided to measure its radio signal.

"From the very beginning, we knew that something was wrong," Wilson recalled. "There was too much noise in spite of the fact that we thought that we had a very good measurement of the gain of the horn. In fact, even before we started working on the horn, we knew that something was wrong."[3] Other Bell scientists who had used the antenna to measure signals from the Echo 1 and Telstar satellites had encountered the same problem. Because of the excess noise, these researchers had failed in their efforts to calibrate the instrument.

In effect, these scientists had simply ignored the extra signal by subtracting it from the measurements, although they used ingenious methods to try to locate the source of the noise. In one instance a Bell scientist named David Hogg beamed a radio signal from a helicopter in an effort to determine whether radio signals propagating through the atmosphere might be having an adverse effect on the antenna.[4]

Throughout 1964, Penzias and Wilson continued their investigation of "Cass A." They hoped that the unwanted noise would somehow simply disappear. Of course, it did not. "We weren't terribly frustrated by the excess noise," Wilson recalled. "It was mainly a background annoyance. Arno, though, was quite embarrassed by it." Radio astronomers are able to describe the intensity of radio signals as a temperature, and when Penzias and Wilson first analyzed the noise they determined that its temperature seemed to hover around 4° Kelvin.* They recognized, though, that the noise could interfere with their astronomical measurements, and so finally decided they had better get serious about getting rid of it. They took the horn apart, correcting what they considered a "suspicious" shape in its throat, they looked for seasonal variations in the excess signal, they replaced rivets they thought might be making an unwanted contribution, they checked for variations related to the phases of the Moon. After all this, they were surprised to find that the excess noise had dropped only to about 3.5° Kelvin.

They went back to work and removed what Penzias has since enjoyed referring to as "white dielectric material" that had been deposited by pigeons within the antenna. Then they removed the offending pigeons themselves (which, taken to another Bell facility some miles away, soon returned). They double-checked their antenna to make sure it was not receiving excess energy from New York City, only about 45 miles away. There was a little noise from New York, but only what was to be expected when the antenna was pointed in the direction of the city.

Penzias and Wilson next tested the possibility that their antenna was receiving noise from charged particles that had been deposited in the Van Allen cosmic ray belts around the earth by a high-altitude nuclear test explosion in 1962. They were hopeful for a while. But it eventually proved not to be the case, since the Van Allen signal at the

---

*Absolute zero, the temperature at which every part of a given system is at the lowest energy level permitted by the laws of quantum mechanics, is approximately -273° Celsius. This temperature is the same as 0° on the Kelvin scale, named after William Thompson, Lord Kelvin (1824–1907), the Irish physicist who developed the scale and the concept of absolute zero. Degrees Kelvin are in the same increment as degrees Celsius, so water freezes at 273.15° Kelvin and boils at 373.15° Kelvin.

7-centimeter wavelength Penzias and Wilson were studying diminished over time, while the peculiar noise being picked up by their antenna went on unabated summer and winter, spring and fall.

In December 1964, Penzias attended a meeting of the American Association for the Advancement of Science in Washington, D.C. There he struck up an acquaintanceship with a fellow radio astronomer named Bernard Burke. Three months later, Penzias happened to telephone Burke to discuss other matters. During the conversation Penzias mentioned the problem with the horn antenna. Burke told him about a group of physicists at Princeton University, only about 30 miles away from Crawford Hill, who might have some thoughts about what was going on with his antenna.

Penzias discussed the news with Wilson for a day or two, then he screwed up his courage and put in a telephone call to Princeton.

OSCILLATING COSMOS

Robert H. Dicke was already a Renaissance man of physics. He was highly regarded for his work both as a theorist and as an inventive experimentalist—a combination rare in the 1950s and 1960s and virtually unheard of today. Dicke commanded respect—even awe—throughout the physics and astronomy world. A 1939 graduate of Princeton University, he had helped develop radar using microwave transmitters and receivers while he was at the Radiation Laboratory at MIT during World War II. In 1946, still in Cambridge, he had invented an acutely sensitive radiometer, an instrument used to measure the intensity of microwave radiation.[5] Dicke also had developed a quantum theory for atoms in a cavity, generated ideas about optical pumping of the energy levels of the atoms, and measured the hyperfine interactions in hydrogen. He had developed an alternative to Einstein's theory of gravity, and then tested it experimentally.

Returning to Princeton in the 1950s, Dicke created a theoretical model for an "oscillating universe," one that would alternately expand and contract. According to the hypothesis, the universe is expanding, but it could have contracted prior to the current expansion and it could contract once again in the distant future. When the future contraction was complete, another Big Bang could take place. Dicke speculated that a remnant of the hot, dense stage of the previous Big Bang in the form of some kind of cosmic radiation might still be observable today. One day in the summer of 1964, Dicke decided to see whether he could track down this remnant radiation, and walked down the hall from his office in the Palmer Physics Laboratory to Room 122. Working there were two members of Dicke's group, P. James E. Peebles and David T. Wilkinson.

Tall and rail-thin, Peebles had an earnest and open expression that belied his mastery of the droll put-down often delivered with a devastating grin. He had graduated near the top of his class at the University of Manitoba, where he had switched from engineering to physics, then gone on to Princeton for his graduate work. There he had fallen into Dicke's orbit and after earning his Ph.D. in 1961, stayed on as a postdoctoral fellow with Dicke. Dicke's group met often for informal seminars, from time to time including specially invited guests who it was felt might make a worthwhile contribution. While eating pizza and drinking beer, the group and its mentor would talk over the big questions of the day.

Wilkinson, too, had fallen under Dicke's spell. Well spoken and thoughtful with a sometimes dour face that could appear standoffish, Wilkinson was, unlike Peebles the theorist, inclined toward the experimental side of physics. He had earned his undergraduate and graduate degrees in physics at the University of Michigan and, after lecturing for a year in Ann Arbor, joined the Princeton faculty as an instructor. He had recently written a textbook on electrons and was looking around for a new project the day Dicke came in with his idea.

Writing on a blackboard suspended from a concrete pillar, Dicke explained how he believed that an oscillating universe would have certain boundary conditions dictated by the oscillations. He speculated that the most recent oscillation would have left the universe with a remnant radiation of a few degrees Kelvin which, theoretically, would be detectable on Earth. Peebles was assigned to do the theoretical work on the radiation, and Wilkinson, along with another member of Dicke's group, Peter G. Roll, would build a new radiometer capable of detecting it.

As Dicke explained the problems involved, Wilkinson thought to himself that the idea was "pretty far out" and that a search for cosmic radiation very likely would prove fruitless. Yet he really didn't have anything else to work on at the moment and the experiment seemed as if it would be "pretty easy and fun." He agreed to go ahead.[6] Roll started building the cold load, and Wilkinson began work on the antenna itself on top of a birdcage on the roof of Guyot Hall, a geology laboratory that was the highest spot around. In those days physicists at Princeton and other universities built most of their experimental equipment themselves, instead of assigning the job to the contractor with the lowest bid. Wilkinson rummaged around in military surplus stores on R Street in Philadelphia for electrical equipment left over from the war. At the time, he recalled, he was vaguely aware of the horn antenna at Bell Labs and that there were a few degrees Kelvin not accounted for in the horn's "error budget." But he also was certain that no radio telescope could do the job of finding cosmic remnant radiation. He thought little more about it as he constructed the instrument.

In the meantime, Peebles was busy working out the theoretical details of the "primeval fireball," as the group had dubbed Dicke's speculation. Peebles and the others received virtually no help from their group leader himself. It was Dicke's practice to hand out an assignment and then step aside and await the results. In fact, Dicke had given Peebles so little to go on that, when the group met some weeks into the project to discuss their progress, another physicist named Nick Woolf stood up and demanded to know why Dicke had not even told Peebles and the others about his old work at MIT.

As it happened, Dicke's team in 1946, in reporting on atmospheric radiation measurements taken by his new microwave radiometer, had found "radiation from cosmic matter at the radiometer wavelengths."[7] Dicke's MIT group, which included Robert Beringer, Robert L. Kyhl, and A. B. Vane, reported that the radiation was quite sparse, less than the equivalent of $20°$ Kelvin. None of the group ever followed up the observation. Dicke explained to his Princeton team members nearly twenty years later that at the time he had not thought of the low-level radiation as residue from cosmic oscillations or a Big Bang–like event, but merely as the possible glow from the universe's most distant galaxies. In fact, he admitted, he had actually forgotten about the measurements.

By early 1965, Peebles had made enough progress on the theoretical end to be able to produce a paper in which he estimated that the remnant radiation would have a temperature of no more than $10°$

Kelvin—only half the upper limit Dicke's MIT group had calculated and then forgotten. Peebles submitted the paper to the journal *Physical Review* for publication. About the same time, he included a version of the paper in a lecture he gave to the Applied Physics Laboratory run by Johns Hopkins University. Almost as an afterthought, he mentioned that his Princeton group was preparing to look for the microwave radiation at the level he had predicted.

Here the route to scientific understanding became circuitous, as so often seems to be the case. In the audience at the Applied Physics Laboratory near Baltimore was an old friend of Peebles's named Ken Turner, a radio astronomer at the Carnegie Institution in Washington, D.C. Turner was especially interested in what Peebles had to say about the possibility of a microwave radiation remnant, and Peebles's ideas stayed with him. A short time later Turner went to Puerto Rico, where he had been working at the large radio telescope at Arecibo. On his flight back happened to be a colleague from the Department of Terrestrial Magnetism at Carnegie, Bernard Burke. Turner passed along the main thrust of Peebles's talk to Burke. A few weeks later Arno Penzias happened to telephone Burke on another matter. And a few days after that Penzias, who admittedly was a little nervous about bothering such an eminent physicist with his problem, telephoned Dicke. When the phone rang in his office, Dicke was having a brown bag lunch with Peebles, Wilkinson, and other members of the group.

"Hmm, cold load . . . horn antenna . . . sounds very interesting," Wilkinson heard Dicke say. Then he asked a few questions and set up a meeting. The conversation lasted a few more minutes, with Wilkinson and Peebles by now sitting at attention. Dicke turned to them as he put down the receiver.

"Boys, we've been scooped," he said, the only time Wilkinson or Peebles had ever heard Dicke use the word. Wilkinson recalled wondering, as Dicke went over his conversation with Penzias, whether the two Bell Labs scientists had taken all the right steps to correct the noise problem in their antenna. Perhaps this was merely wishful thinking on Wilkinson's part. For he would soon learn that Penzias and Wilson had done everything necessary to bring about a major change in the way he, Dicke, Peebles, and everybody else would think of the universe.

AN AGE OF INNOCENCE

Late in the winter of 1964–65, I was at Swarthmore College, not far from Bell Labs and Princeton. As a freshman physics student, I was happily oblivious to the momentous struggle of Penzias and Wilson to overcome the irritating noise in their antenna and to the problems in cosmology being worked out by Dicke and his group at Princeton. My major concerns were trying not to flunk art history and not to be bored in first-year calculus and physics. Fortunately, I had been allowed to skip the second half of my freshman physics course and go straight into sophomore physics.

As long as I can remember, I had thought science was the most interesting pursuit possible. My parents first thought I would grow up to become a scientist the day I took off the doorknobs in the house when I was three years old. When I was about five, my father would sit by my bedside and explain how living things are constructed of minuscule cells with nuclei and chromosomes that in mysterious ways could control our inheritance and thus our future. He was a geneticist and spent his life trying to breed dairy cows that would give milk with higher protein and with greater efficiency. He had been born in a tiny village in Africa, and spent his first nine years in what was then Southern Rhodesia. He had many stories to tell. I remember asking how he had decided to go into dairy cattle research—not exactly a glamorous field. When he had been growing up in Africa, he explained to me, he had looked at the small, scrawny cattle that lived there compared with the big, healthy ones he saw in the dairy magazines his father received from the United States. He decided to see whether he could do something about it.

Unlike my father, I was a child of the cold war. I was born a year and a day after the Enola Gay dropped an atomic bomb on Hiroshima. When I was in fourth and fifth grades, we were trained to hide under our desks for protection from a possible hydrogen bomb attack. As one of the few scientists in the county, my father was a member of the civil defense program and understood how a Geiger counter worked. When the first Soviet satellite, Sputnik, was rocketed into orbit in 1957, I can remember being thrilled and wondering whether I should be terrified too. In the ensuing panic over our nation's apparent loss in science leadership, there was renewed emphasis on science education. Science fairs sprang up across the country on the theory that our young scientists were now the only ones who would be able to save the nation from the Soviet threat. Such an atmosphere, of course, could not help but be beneficial to a budding young scientist such as myself. In reality, of course, I was far from the fray.

My family lived on the Dairy Research Station of Rutgers University near Sussex, New Jersey. It was a rather idyllic life in that my father simply left in the morning for an adjacent building where he worked all day with his test tubes and calculator, a giant old Marchant or Friden machine with a couple of hundred buttons on it. The farm country in New Jersey was quite lovely, with rolling hills and dales. There were said to be more cows than people in the county until the 1950s, and the Appalachian Trail ran along the top of a ridge near our house. My mother was a teacher at the Wantage Consolidated Elementary School, where I was then a student. A dedicated teacher, she made sure all her first-grade students knew how to read and write a bit by the time they finished their year with her—more than can be said of many high school students in the 1990s. She had an extremely positive influence on me, encouraging me to do what I wanted to do and what I believed I could do best.

That was astronomy, which always had intrigued and excited me. In 1953, when I was seven, my parents took my sister Janet and me to visit the Museum of Natural History in New York City. At the time Mars was at opposition and unusually large. The newspapers were full of stories about how the new Palomar telescope in California would be able to observe the Martian canals. People were hoping for space aliens, as they still do. At the museum I was deeply impressed with the planetarium show, the orrery (a giant working model of the solar system), and a huge iron meteorite weighing several tons that was thought to be a fragment of the missing planet between Mars' and Jupiter's orbits before it struck Earth.

The museum's fine dinosaur exhibit contained displays of the evolutionary history of various species from millions of years in the past. It was beautifully clear that my father's stories of genetics and evolution were right. My parents already had read to me from biographies of Galileo and Darwin, and I thought science could be a dangerous but glorious adventure. I remember having bad dreams about the Scopes "monkey" trial, which had taken place a generation earlier, and imagined myself being persecuted for teaching evolution. I loved *The Microbe Hunters* by Paul DeKruif, with people battling deadly germs that could bring them a slow, lingering death; my interest undoubtedly was kindled by the fact that my maternal grandfather, Hobart Cromwell, had helped develop penicillin at Abbott Laboratories.

Since early childhood I loved to play with all kinds of instruments and measuring devices. In third grade I was given a radio kit with a single vacuum tube laid out on a board with clips and leads. I built the radio, even though my receiver would not be able to pull in even a single station, since we lived too far from town. When I was about ten, our county library began a bookmobile service to our farm and others. I still think of the bookmobile, my link to the outside world, as a magnificent institution for discovery. I read everything I could find on astronomy and electronics. My favorite was the Amateur Telescope Maker, a series of three books from *Scientific American* that I read over and over until I had almost memorized them.

While other kids were collecting baseball cards, I was reading science books. These confirmed what I already had begun thinking: The most glorious thing you could do in science was to build a new instrument such as a telescope that could see something no one had ever seen before, and the people who built such instruments were my heroes. I memorized the pages of the Edmund Scientific and the Allied Electronics catalogs, and spent days tinkering in the basement with bits of wood and metal. By the time I entered high school, in 1960, I had become what is now called a nerd. But I still think it was wonderful to have had the opportunity to develop the skill for these things.

During high school summers I got into National Science Foundation programs, attending Assumption College to study mathematics one year and Cornell University to study physics another year. I also managed to save enough allowance money to buy the parts to build a telescope, a little 4-inch reflector. I applied to my parents for a research grant and for $25, a lot in those days, I purchased a war-surplus aerial camera lens. I planned to use the telescope and the camera to compute the orbits of the planets, as Johann Kepler and Karl Gauss had done. This proved much more difficult than I had imagined, and my lack of success certainly raised my admiration for the science pioneers who had accomplished so much with so little. I saw that an understanding of calculus would be necessary in order to calculate the planetary orbits, and I signed up for a study hall where I learned calculus from the same book my father had used.

Our life on the dairy research station was pastoral and peaceful. I rode an hour each way to school on a yellow bus. I had no social life. What would it have been like to attend a big-city high school with other kids who wanted to be scientists? I often wondered. I had my successes, winning the statewide physics contest in 1963 and placing in the top ten in the New Jersey high school mathematics contest. But my parents warned me that so far I'd just been a big fish in a little pond and the world outside could be far more challenging. I yearned to go away to college to have more friends. When the time came I chose Swarthmore College near Philadelphia over Harvard and MIT because it had an excellent academic reputation and was located in a quiet sub-urban atmosphere fairly close to my home in New Jersey.

By chance I would meet there one of the men involved in the scientific intrigue that took place at Bell Labs and Princeton University in 1965. This was Dave Wilkinson, with whom I would work on the COBE project for twenty years.

☀ 3 ☀

The Remains
of the First Day

I feel quite certain he really did see an apple fall.

—PROFESSOR RUPERT HALL, NEWTON SCHOLAR

Jim Peebles's paper detailing a possible theoretical interpretation of the universe's afterbirth in the form of remnant radiation was rejected for publication in early 1965. Peebles was nonplussed by a cryptic review letter accompanying the rejection note from the editor of the journal *Physical Review.* The letter, supplied anonymously, as is customary, stated that, while the paper was well written and its physics first-rate, the author's scholarship was poor. In effect, it said, 90 percent of the work in the paper had been done years before. Moreover, Peebles had failed to attribute the earlier work. The note listed a number of references the author was advised to consult. Peebles revised the paper and sent it back to the journal, but it was rejected a second time for what amounted to the same reasons. A subsequent version of the paper was eventually published in another journal the following year.

This was the preamble to what would prove to be the most unfortunate part of the story of the discovery of the cosmic background radiation. Apparently unknown to either the Bell Labs scientists or the Princeton physicists, a third group of researchers had completed seventeen years earlier a major portion of the theoretical work that Peebles had undertaken from scratch. In the late 1940s George Gamow, a Russian-born physicist then teaching at George Washington University, and two young colleagues, Ralph Alpher and Robert Herman, had begun investigating what would come to be called the Big Bang model for the creation of the universe. Although not adept at everyday matters such as spelling or even mathematics, Gamow possessed an unusual genius for asking penetrating questions and broaching new ideas for his colleagues to pursue.

Born in Odessa in 1904, Gamow had studied at the University of Petrograd in St. Petersburg during the 1920s with the theorist Alexander Friedmann, who had speculated mathematically on the ways the universe might evolve in brilliant new solutions to Einstein's field equations of general relativity. Later Gamow moved on to the university in Göttingen, Germany, where he performed so well in particle physics that he drew the attention of Niels Bohr. Bohr was a Danish theorist who by then was legendary for his role in the creation of quantum mechanics, the branch of physics devoted to the fundamental particles of matter and the forces controlling them. In 1928 Gamow formulated an early theory about radioactive decay and was one of the first physicists to address the problem of how stars evolved.

In the 1940s Gamow, then living in the United States, became well-known for his popularizations of physics and astronomy such as *Mr. Tompkins in Wonderland* and *One, Two, Three ... Infinity*—books that inspired many a young astronomer- or physicist-to-be, including me. An interesting and vociferous individual, Gamow was an inveterate practical joker and, unfortunately for his role in the events that were to follow, an alcoholic.

In 1944 a part-time evening graduate student at George Washington named Ralph A. Alpher became interested in the ways subatomic particles might have been created in the early stages of an expanding universe. This was Gamow's field. Alpher went to him and asked him to serve as his Ph.D. adviser. Gamow agreed.

Times were tough right after the war. During the days, Alpher worked at Johns Hopkins's Applied Physics Laboratory near Baltimore to pay for his graduate classes. He had met there another young physicist named Robert Herman, who had recently received his doctorate from Princeton. Alpher and Herman became fast friends at work.

Inevitably, since their interests coincided, they began collaborating. Soon, as the three shared a strong fascination with the mechanics of the creation of the universe and the evolution of stars, Alpher and Herman began working with Gamow. He would serve as the two young theorists' mentor for the remainder of the decade.

CAT'S CRADLE

The collaboration of Alpher, Gamow, and Herman in the 1940s occurred during a period when astronomical observation and theory were beginning to merge in the new science of cosmology. Theory, the effort to guess and understand the universal laws of nature, and observation, the effort to test those guesses, were the engines that, in often uneasy tandem, drove the machine of science. Isaac Newton had achieved the first significant pairing of theory and observation in the seventeenth century in his mathematical description of the force of gravity.

Newton's Law quite literally transformed science. For the first time the astonishing powers of mathematical models of the natural world could be seen, at least by the few who could master the methods. Without Newton's Law the development of cosmology simply would not have been possible. It was a magnificent coupling of Johannes Kepler's observational laws relating to the solar system—particularly Mars, the most difficult of the planets because of its eccentricity— with the very accurate astronomical data of Tycho Brahe. A Danish nobleman who had given up the traditional pursuits of his class for astronomy, Brahe may have been the most generously supported astrophysicist of all time: The king of Denmark granted him the island of Hven and all its income for life. This was a significant portion of the kingdom's gross national product, and Brahe used it to build a remarkable observatory, Uraniborg.

Presiding there like a feudal lord for twenty years, he and his assistants used quadrants with radii as great as 14 feet to bring the science of celestial observation to the greatest accuracy possible with the unaided human eye. After a falling out with the king, Brahe left Denmark for Prague in 1600, taking his astronomical records with him. There he met Kepler, a Protestant who had recently fled his post at the

Catholic University of Gratz in Styria before a wave of religious hostility. Brahe died the following year of an infection that resulted from the bursting of his bladder during an audience with the emperor. Too nearsighted to observe the stars and planets himself, Kepler managed to appropriate Brahe's observational data from his heirs.[1] Kepler then combined these data with his own theoretical insight and intuitive feel for the mechanical, incorporating newly available mathematics texts from the ancient Greeks about ellipses. He concluded that the Sun possessed a power, an *anima motrix,* that caused motion in the planets. Seeking an explanation of celestial dynamics that went beyond a mere law of motion, Kepler teetered for years on the brink of a law of accelerating force. His undoing was that the power he invested in the Sun decreased at a linear rate with distance rather than as the inverse square of the distance.[2]

Such an insight was to be the triumph of Newtonian cosmology. Newton apparently really did start thinking about gravity after seeing an apple fall.* Then, during an eighteen-month period before he turned twenty-four, Newton worked out the laws of motion and universal gravitation, showing that the force that made the apple fall to Earth and the force that kept the Moon circling in orbit were one. In contemporary terms, he had discovered an underlying symmetry between two seemingly dissimilar physical interactions. No matter their mass or composition, any two objects attracted each other at a rate directly proportional to the product of their masses and inversely proportional to the square of their distance apart. A mathematical tool also invented by Newton, calculus, explained why the apple fell straight down to the ground rather than, for instance, sideways toward another massive object such as a nearby mountain. For a spherical Earth, the force must be radial, as though the mass is concentrated at the center, and (for a $1/r^2$ force) the distant parts of the earth matter much more than the nearby mountain.

---

*Professor Rupert Hall, a leading Newton scholar, formerly at London's Imperial College, has stated that he is sure Newton really did study a falling apple: "Voltaire first reported it, and Newton himself confirmed it at least twice."[3] Western history's second most famous fruit tree, or at least a clone grafted onto the stump of the original, which was felled by lightning, still stands at Woolsthorpe Manor, Newton's birthplace. The tree is the only known source of a pinched, foul-tasting little seventeenth-century apple called the Flower of Kent.

Fearing that his ideas would be appropriated if not outright stolen by others, a practice common at the time, Newton refrained from revealing his findings for more than two decades. And only by accident was his work published even then. Edmund Halley, then Astronomer Royal, was unable to calculate the orbit of a comet that appeared about every seventy-five years and went to Newton for help. Newton told the astonished Halley that he already had done the calculation as part of the derivation of his new law of gravitation. When he was unable to find it among his papers, he did the calculation again on the spot. Halley recognized at once Newton's remarkable accomplishment, and offered to pay for its publication.

In the *Principia,* which finally appeared in 1687, Newton unified physics and astronomy as a single science of matter in motion. Moreover, he established the modern balance between physics as the science of metrics and mathematics as the language of quantity.[4] Gravitation itself was a sort of cosmic cat's cradle of forces, with every star and planet (Newton was unaware of galaxies) tugging at every other celestial object across the great void of the cosmos. With his new force, Newton managed also to reconcile the continuity of space with the discontinuity of matter—an apparent dichotomy that was the last of the Greek philosophical problems made irrelevant by Western European science.

It has been suggested that Newton himself may have contemplated the consequences of an infinite space filled with matter acting upon itself gravitationally and realized that such a condition was inherently unstable. Had he or another seventeenth-century scientific philosopher followed such a line of reasoning to its conclusion, it might have been possible even at the time to suppose that the universe was expanding. Newton was not likely to have participated in such conjecture. Writing in the "General Scholium" for the second edition of the *Principia* in 1713, when he had grown old but not mellow, Newton attacked his critics for drawing from his laws of nature such kinds of hypotheses, which he wrote "have no place in experimental philosophy."

Nor would Newton speculate about the cause of gravity for the same reason. This would be left to Albert Einstein, who two centuries later would devise the theory of special relativity from a simple principle: The laws of physics should be independent of the steady linear motion of an observer. The insight led Einstein to conclude that the speed of light must also be a constant. Moreover, its velocity could be deduced from the equations of James Clerk Maxwell, the Scottish physicist who had shown that electricity and magnetism are manifestations of the same force, whose constants could be measured with stationary apparatus.

Einstein realized there was only one way his new principle could be implemented: The Newtonian universe had to be overthrown. Space and time would no longer be understood as independent and absolute. Before Einstein, scientists thought they knew how to compute the physical effects seen by a moving observer. All observers would be able to read the same clocks, so there was no problem. An experimenter in a moving train would see the world moving by, and an experimenter by the tracks would see the train moving by. Since it was thought that both time and space were absolute, the calculation was easy. Einstein noticed that in the old way of thinking, the observer on the train could measure the velocity of light forward and backward, and should get different answers because the train was moving relative to absolute space. On the other hand, how does the observer in the train know his velocity in absolute space if the windows are closed and the curtains drawn?

It seemed more appealing, more likely, and more believable to Einstein that all observers, whether moving at a constant velocity or not, should measure the same values of the constants in Maxwell's equations, and therefore compute the same velocity of light. There was only one escape. All observers cannot be using the same clocks, and neither time nor space can be absolute. All we can do is measure in relative terms. To modern physicists who grew up on childhood stories of Einstein, such an approach seems perfectly natural. But to the majority of scientists ninety years ago, it was wholly unacceptable. Einstein was denounced for it, even though his new concept explained the Michelson-Morley experiments that had proved nearly two decades earlier that the velocity of light was a constant, regardless of the steady motion of the observer. When he finally received the Nobel Prize, it was for entirely different work on the photoelectric effect. The theory has many consequences, including the famous equivalence of mass and energy $E = mc^2$, and by now it has been tested with extraordinary accuracy.

Although such heresy was initially opposed bitterly by many physicists, the beauty of such an idea and its descriptive successes eventually won the day. Moving on, Einstein wanted to create a theory of gravity consistent with special relativity but that eliminated the need for inertial frames of reference. Recognizing that Newton's gravity acted equally on all matter, Einstein showed that gravity does its work by bending the very geometry of the space-time in which the matter moves.

In this picture he did not think of the earth as following a curved path around the Sun. Instead, he thought of the planet as following a straight line (or, in geometric terms, a geodesic—the shortest curve between two points) within his new four-dimensional curved space-time. Fortunately, Nikolai Ivanovich Lobachevsky and Georg Friedrich Bernhard Riemann, nineteenth-century pioneers in non-Euclidean geometry, had already laid the groundwork and created the mathematical tools Einstein needed to describe curved space-time. Einstein found there was only one natural way to do such a calculation. In modern notation, this was the equation $R = T$, where R is the curvature tensor describing the shape of the space-time and T is the stress-energy tensor that includes terms within it to describe the motion of matter and energy, both of which can produce gravity.*

Einstein's new theory of gravity, general relativity, was published in 1915, and was tested almost immediately for two famous predictions it made.[5] General relativity predicted that the motion of Mercury's ellipse was precessing about 39 arcseconds more per century than Newton's law of gravity had predicted. This had been a known problem for decades, so Einstein's success was immediate. His new theory also predicted that light passing near the Sun would be bent in such a way that the positions of stars would appear to shift just as the Sun passed in front of them. Although such an effect ought to occur even under Newton's laws, Einstein predicted a value twice as large as that of the old theory.

Arthur Stanley Eddington, director of the observatory at Cambridge University and one of the first physicists to understand the full significance of general relativity, organized an expedition to view a total solar eclipse in 1919. His observations confirmed that light rays, when subjected to a strong source of gravity such as the Sun, bend precisely the amount predicted by general relativity. Soon other scientists (and eventually everyone) would realize that the universe was a far more complex and fascinating place than they had imagined.

---

*The famous equation $E = mc^2$ shows that matter and energy can be freely interconverted, and that therefore they should both be sources of gravity as well.

One was Alexander Aleksandrovich Friedmann, the University of Petrograd theorist who served briefly as George Gamow's mentor. Friedmann, a sad-looking man with a wan face and drooping mustache, was that rarest of beings, an intellectual with no regard for the practical implications of his theoretical investigations. He had been trained as a mathematician and meteorologist and possessed an unusually versatile intellect. He decided to apply Einstein's field equations of general relativity to the entire cosmos. Working without the burden of philosophical or political restraint in turbulent postrevolutionary Russia, Friedmann soon discovered that the universe could take any of several forms if he recalculated the equations of general relativity.[6]

In groundbreaking papers in 1922 and 1924, Friedmann demonstrated mathematically that the universe could very well be a dynamic system that, regulated by the gravity of general relativity, could expand indefinitely or collapse back on itself like a deflated balloon.[7] A third possibility was that the universe was in a state of precise balance between infinite expansion and collapse. What would determine the true dynamic of the universe? According to Friedmann, the average density of mass within the universe would define how space curved as described by general relativity. Such a curvature would establish the way the universe changed over time.

Einstein would have none of this. Philosophically insecure with anything but a static universe, he had inserted into the equations of general relativity his famous "cosmological constant." This was a mathematical contrivance aimed at preventing just the kinds of unstable universe predicted by Friedmann who, in making a number of simplifying assumptions, had removed the constant from his own mathematical calculations.*

THE GROWING UNIVERSE

Theory alone could not have produced modern cosmology. Without real data from the real universe, the theoretical ideas of Newton, Einstein, and Friedmann would have been little more than interesting spec-

---

*It was later shown theoretically that Einstein's universe with its cosmological constant is also unstable and ought eventually to collapse or blow up unless other factors—still unknown—are at work.

ulation. In fact, that is precisely all that Friedmann intended. Historians of science today customarily trace the concept of an expanding universe to Friedmann's 1922 and 1924 papers. Yet Friedmann had concluded only that nonstatic mathematical solutions to the field equations of general relativity were *possible*. He attached no physical significance to his equations, nor did he try to connect his results to actual astronomical observations.[8]

By a coincidence of events, astonishing new details about the universe soon appeared. In retrospect, these made Friedmann seem highly prescient. During the nineteenth century most large reflecting telescopes were built in Ireland and England, usually funded by enthusiastic and wealthy amateur astronomers like William Parsons, Lord Rosse, who built the "Leviathan of Parsonstown." This was a 72-inch reflecting telescope at Birr Castle in Ireland that identified the first spiral nebula, in 1845.[9] During the years before and after World War II, the United States rose as one of the world's leading powers. A side benefit of the nation's new economic might was that American astronomers were able to persuade wealthy businessmen to pay for expensive and powerful new telescopes to be placed on good sites under the clear skies of the western part of the nation. The greatest of these was the huge 100-inch Hooker reflector installed on Mount Wilson near Pasadena, California, in 1917.

The Hooker was put to its finest use by Edwin P. Hubble during the 1920s. Born in 1889 in Missouri, Hubble had moved with his family to the suburbs of Chicago, where he eventually won a scholarship to the University of Chicago. Handsome and robust, he was adept at whatever he tried. As a young man he had been a good boxer and was successful enough academically to earn a Rhodes scholarship. During his three years at Oxford University, he studied Spanish and jurisprudence, and earned a law degree. He also became intrigued with astronomy and, a year after returning to the United States, enrolled as a graduate student at the Yerkes Observatory of the University of Chicago.

After completing his dissertation, "Photographic Investigations of Faint Nebulae," and serving with the American Expeditionary Force in Europe, Hubble joined the staff of the Mount Wilson Observatory near Pasadena in 1919 and began making observations with the Hooker reflector. In the early 1920s Hubble aimed the telescope at a fuzzy white spot known as the Great Nebula in the Andromeda constellation. Along with most other astronomers, he assumed this tiny cluster of light, visible to the naked eye on a clear night, was a glowing cloud of gas—a nebula—floating between the stars of the Milky Way.

Hoping to establish its distance, Hubble took dozens of photographic plates of Andromeda. In October 1923 he made a dramatic find when, during a single week, two stars appeared on his plates. He first thought they were novae, a kind of star that increases suddenly in brightness and that previously had been seen in Andromeda. Because of the uncertainties of their identification and difficulties in measuring their absolute brightness, novae were of almost no use as distance indicators. Invariably thorough, Hubble checked earlier plates and drew a light curve—brightness versus time. He realized at once that the new objects were not novae but another kind of star, a Cepheid variable, whose brightness changed over time in a cyclic and highly predictable manner.

By calculating the period of one of the Andromeda Cepheids, Hubble could determine the star's true brightness by relying on an established relationship between the periods of other known Cepheids and their true luminosity.* Assuming that Cepheids everywhere have the same absolute luminosity compared with their period of light variation, he could compare the observed brightness of his new Cepheid with its actual brightness. A quick calculation gave him an estimate of the distance to the Andromeda nebula that contained the Cepheid: one million light-years. He found other Cepheids that confirmed the results. Within a year he had accumulated enough evidence to convince other astronomers that Andromeda was no nebula. It was a galaxy made of stars like our own Milky Way, but suspended at what to astronomers and everybody else in the 1920s was an unimaginable distance out in the vastness of space.

By careful and convincing work with the Andromeda Cepheids, Hubble settled once and for all a long-running debate among astronomers over the true nature of nebulae. The term *universe* also

---

*The relationship between the period and absolute luminosity of Cepheids was discovered in 1908 by Henrietta Swan Leavitt of Harvard University while studying stars of fluctuating brightness in the southern group called the Small Magellanic Cloud. Carefully examining sixteen stars whose light variations she had measured, she found that the longer its period of fluctuation, the brighter the star.

took on a new meaning. To early twentieth-century astronomers, the word meant only our galaxy, the Milky Way. When the English physicist and astronomer Arthur Stanley Eddington wrote his well-known book *Stellar Movements and the Structure of the Universe* in 1914, his title addressed the dynamics of stars and the structure of our home galaxy.[10] After Hubble, astronomers realized that the universe we inhabit is vaster by many, many magnitudes than anyone had ever dreamed. That space might be infinite was no surprise—the Greeks had deduced that more than 2,500 years ago—but no one before Hubble had ever known what filled its farther reaches.

During the next five years Hubble made an even more astonishing discovery: Not only was the material universe far bigger than anyone had suspected, but it was growing at a startling rate. The groundwork for Hubble had been lain by Vesto Melvin Slipher, an astronomer at the Lowell Observatory near Flagstaff, Arizona. Around 1909 Slipher began aiming his spectrograph, which he had attached to a 24-inch refracting telescope, at Andromeda and other nebula. By 1912 he had produced four photographic plates that greatly puzzled him: The spectral lines of Andromeda were discernibly shifted.

Slipher decided such a displacement must be caused by a Doppler shift, which could only mean that Andromeda was moving rapidly. He made a quick calculation and determined that the nebula was traveling about 300 kilometers per second, the greatest speed ever recorded for an astronomical body at that time.[11] In August 1914 he announced to a stunned meeting of the American Astronomical Society (attended by Hubble, then a graduate student at Yerkes Observatory) that most of the fifteen nebulae he had measured were traveling away from our solar system at speeds of about 1,100 kilometers per second. A leading astronomer, William Wallace Campbell, told Slipher his announcement was one of the greatest surprises astronomers had ever had.[12]

But nobody knew what it meant. The nebulae were moving faster than the fastest stars in the Milky Way by several hundred kilometers per second. Slipher's discovery implied that the nebulae could not be bound gravitationally to the Milky Way. But this was impossible, since most astronomers thought the nebulae were part of the Milky Way. Ten years later Hubble and his assistant, a remarkable man named Milton Humason who was a former mule driver and self-taught astronomer, began measuring the velocities of the most distant galaxies the Hooker telescope could discern. Slipher had been right. By a careful analysis of their plates, Hubble and Humason saw that light from the far galaxies was changed by the time it reached Earth. As Slipher had already determined, the wavelengths of the light from the galaxies had shifted toward the red end of the spectrum.

The redshifts indicated to Hubble, as they had to Slipher, that the galaxies were receding from the Milky Way at enormous speeds. But what did it mean—what was the big picture? By 1927 astronomers had speculated for more than a decade about the possibility of a relationship between a galaxy's velocity as determined by its redshift and its distance. But the data were not convincing. The plots of redshift compared with distance looked entirely random. Hubble studied and restudied these diagrams. One day in 1929 he suddenly realized that there was a pattern after all: The more distant a galaxy, the faster it appeared to be receding. He soon discovered that the velocities of the galaxies increased in a direct relationship with their distance: One twice as far away as another had twice the speed of recession, while one four times away traveled four times as fast.[13] The simplest explanation of this relationship, which has since become known as Hubble's Law, is that not only is the entire universe not static; it is expanding rapidly. Friedmann had been right. But nobody, including Einstein, had recognized this at the time. Friedmann had died in 1925, his papers largely unread.

It would be left to one eccentric individual to show that Friedmann's work had been ahead of its time: a Belgian abbé named Georges Henri Joseph Edouard Lemaître, who was a professor of mathematics at the University of Louvain. Lemaître independently derived Friedmann's equations before Hubble's discovery, and was the first person to bring about a marriage of the theory of general relativity with Slipher's and Hubble's astronomical observations about the universe.

* 4 *

The Day the
Universe Changed

Your calculations are correct, but your physical insight is abominable.

—ALBERT EINSTEIN TO GEORGES LEMAÎTRE REGARDING HIS
THEORY OF AN EXPANDING UNIVERSE, 1927

IF THE UNIVERSE WERE expanding, the question remained: What had it expanded from? Georges Lemaître, one of the strangest characters to wander onto the stage of twentieth-century physics, was the first one to attempt an answer. Born in Belgium in 1894, Lemaître was plump, irritating, and ahead of his time. In 1927, unaware of Alexander Friedmann's work, Lemaître published a paper in an obscure Belgian journal in which he drew a mathematical theory that linked general relativity with the comparatively few redshifts that already had been seen. Lemaître concluded in the paper that the universe *must* be expanding.[1] His hypothesis was two years before Hubble's announcement that he had discovered galaxies in recession.

Later the same year at the fifth Solvay conference on physics in Brussels, Lemaître tried to get Einstein's attention. Normally tolerant and kind, Einstein pushed him aside abruptly, saying, "Vos calculs sont corrects, mais votre physique est abominable."[2] Undeterred by Einstein, already the most famous physicist in the world, and bolstered by the confirmation Hubble's redshift observations had given his new theory, Lemaître extrapolated his theory to what seemed to him its logical conclusion: The universe *must* have originated in a primordial explosion.

A letter Lemaître wrote to *Nature* magazine in 1931 was effectively the charter of what was to become the Big Bang theory. He theorized that this primordial explosion, occurring on "a day without yesterday," had burst forth from an extremely dense point of space and time. He began calling this the "primeval atom."[3] By now Lemaître had become a celebrity in his own right for his revolutionary ideas. At an immense gathering of the British Association for the Advancement of Science in London the same year, he speculated before an audience of several thousand scientists that cosmic rays may have originated in the primordial explosion. Eventually, he thought, they might prove to be material evidence of the universe's "natural beginning."*

During the years before World War II, cosmology divided into two camps. One camp, the historian John North observed, "argued from geometrical and kinematic premises. They tended (for example, in the appeals to symmetry) to a rigid idealization of the situation. Often this was not accompanied by any but the slightest reference to astronomical practice."[4] The other camp started from the question Einstein had raised about the stability of the universe, then undertook a serious calculation of Lemaître's primeval atom. By the late 1930s this second camp had entered into an exchange between theory and observation that would have been thought impossible just a decade earlier (although minuscule by today's standards). What accounted for this new exchange of information between the occasionally warring branches? For the first time each side was dependent on the other, since the observations of Hubble were being coupled with the revolutionary theories of Lemaître.

---

* In an unpublished manuscript written about 1922, Lemaître wrote that "as the genesis suggested it, the Universe had begun by light." He acknowledged that this was based only on intuition, and never followed through with the idea for lack of a scientific basis. Lemaître, who grew old watching more celebrated physicists follow the course he had set, learned on his deathbed in 1966 of the discovery of the cosmic background radiation, which suggested that the path he had started down had been the right one.

THE BLACKBODY UNIVERSE

Lemaître's ideas were tentative and speculative. In the 1940s George Gamow and his two young colleagues, Ralph Alpher and Robert Herman, began to try their hand at refining the theory of the primeval atom. Essential to their work was the concept of blackbody radiation. Its theoretical basis was developed shortly after the turn of the century by Einstein and Max Planck, the famous German physicist who was one of the founders of quantum mechanics. Blackbody physics derives its name from the theorized property that a black body absorbs all incoming radiation, then reradiates it in a highly characteristic pattern first predicted by Planck.

In a closed box with completely opaque walls—a "blackbody cavity"—the walls (as would any other object made up partially of electrons) emit electromagnetic radiation. This can also be thought of as "radio noise." An idealized blackbody is a large box blackened inside. If you hoped to see inside, you could poke a very small hole in one wall. Light going into the box through the hole would have almost no chance of coming out again. As you peer in, everything would be perfectly black. The hole would be so small as to have virtually no effect on radiation going in or coming out. This means that you could use the little hole to measure the temperature inside without having a significant effect on it. Remarkably, the intensity of the radiation in the box would depend only on the temperature of the walls and nothing else: the higher the temperature, the greater the radiation's intensity. This simplicity is a result of the fact that the interior of the box is at equilibrium with itself. No part is warmer or colder than another. The light waves at each wavelength are in equilibrium with those at every other wavelength, as well as with the walls of the box. Planck discovered that if he plotted the intensity versus the product of wavelength and temperature on a graph, he would always end up with the same shape of curve (see figure 1). He invented a simple formula to describe the experimental data, and it turned out to be exactly right according to quantum mechanical theory.

By the late 1940s, Gamow, Alpher, and Herman were thinking seriously about the primordial explosion. Gamow had made a customarily daring theoretical gambit, asking whether it would have been possible for the chemical elements to have come into existence before the creation of the stars themselves. The Big Bang, as it happened, was

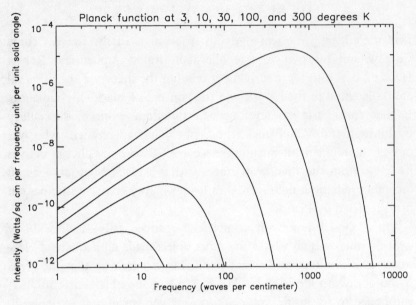

Fig. 1. PLANCK FUNCTION.
This graph depicts blackbody radiation intensity versus frequency. Planck showed that all these curves have the same shape.

an obvious place to look. Inconceivably dense, it was a giant pressure cooker with plenty of energy to make things happen. It was already understood that stars could cook hydrogen into helium, liberating the energy that would fuel them and, eventually, warm the earth, but it wasn't yet known what made the other heavier elements.

It was not Gamow's style, or his interest, to work out the mathematical details of his theoretical ideas and speculations. With his characteristic enthusiasm he had no difficulty persuading Alpher to work out the nuclear synthesis for his dissertation. Alpher worked at his day job at the Applied Physics Lab and did his thesis at night, collaborating with Robert Herman. Their strategy was to work backward from the present, which as far as they knew was a universe filled with stars, made mostly of hydrogen with about a quarter helium, and a little starlight. Somehow they had to calculate the conditions at very early times. They knew that the density and temperature in the early universe were so high that particles and antiparticles could be quickly created from pure energy and then annihilated a moment later. In a way, that simplified the problem enormously.

A little imaginary box of the primordial soup would almost instantly achieve a kind of thermal equilibrium, and they could calculate its properties just from knowing the temperature of the soup. Nothing would be solid; the universe would be a giant nuclear pressure cooker containing a seething froth of light and particles. Of course, as the soup and the imaginary box expanded they would cool, but that happened very slowly compared to the time it took to cook the particles to equilibrium. They did not know the temperature, so they had to work out what would happen for a wide range of possibilities.

They knew this situation could not last forever, because the universe has since moved very far from equilibrium. They calculated the details for a time a few minutes after the great explosion when the material was primarily composed of neutrons, protons, electrons, positrons, photons, and neutrinos. With a half-life of about fifteen minutes, the neutrons would decay or react with the protons to build up heavier nuclei. Then the neutrons would be gone and the remaining nuclei would be too cold to make nuclear reactions. Although they were initially motivated to explain the formation of all the chemical elements, they found a surprise: Nothing but the two simplest elements, hydrogen and helium, could be formed. There would be only extremely small traces of anything heavier.

In a sense, Alpher and Herman had failed. Gamow's initial guess was incomplete. The full story of the chemical elements would be left to others to tell. It would turn out to be extraordinarily elaborate, involving exploding stars and intricate networks of nuclear reactions that were only just becoming understood in the 1940s. The legacy of the war had brought us the knowledge and the tools to answer the question of how our own chemical elements got here so we might exist. Alpher and Herman published their results in 1948,[5] Alpher completed his dissertation and published it in the same year,[6] and then they published yet another paper that year with Gamow on the primordial nucleosynthesis.[7]

THE TWO-BILLION-YEAR-OLD FOOTPRINT

The most remarkable paper in that remarkable year for Alpher and Herman had little to do with formation of the elements. For their work on nucleosynthesis for Gamow, Alpher and Herman had modeled the first moments of the universe. As the details emerged, they started realizing that the radiation released in the primordial explosion should have maintained its blackbody pattern throughout the entire history of the universe—right down to the present day.

Alpher and Herman had taken an incredible step. Had it been recognized at once by the astronomy community, the course of cosmology for the rest of the century would have been far different—as would have been the lives of Alpher and Herman. They were the first to recognize that even as the expanding universe cooled and attenuated the primordial radiation, its essential pattern would remain the same. This was roughly analogous to what might have happened to the footprint of an early hominid implanted in African mud a million years ago. It would change, of course, as it was filled and refilled with water and sand over the years, but its essential shape could still remain today.

Alpher and Herman believed this was the case for the radiation from the beginning of time. It still had a characteristic blackbody pattern. Could they give the blackbody a number? They could because this was, in fact, the easy part of the job. They simply looked at existing estimates of the density of matter in the universe today and compared it with their estimate of the density of the universe at the moment of the primordial explosion. A fairly quick calculation revealed what the temperature of the blackbody radiation should be at the present time. This was, Alpher and Herman figured, approximately 5° Kelvin.[8] They published their calculation in the prestigious British journal *Nature,* then refined their mathematics and published their prediction of the cosmic background temperature twice more during the next three years.

Their prediction was virtually ignored for nearly two decades. Nobody knows exactly why, although a number of theories have been put forth. Several factors undoubtedly contributed to the lack of interest in what was to prove a remarkably prescient prediction. A technical reason may have been that astronomers at that time may simply have believed, like Robert Dicke, that they did not know how to measure cosmic background radiation with their existing equipment.

I disagree. Although such a measurement would have been difficult with 1940s and 1950s technology, it was possible. Dicke's team had already tried in the wartime 1940s, but without the motivation to try harder they had given up. All the necessary parts were available. It's just that nobody tried. Another student of Gamow's, Joseph Weber, heard about Alpher and Herman's prediction and asked about the feasibility of measuring the primal radiation. Told it was impossible, Weber gave up the idea and went on to the University of Maryland, where he led the development of a technology for measuring gravitational waves predicted by general relativity.

Alpher and Herman's association with Gamow may also have been a negative factor. Gamow was recognized as extremely creative by the astronomy community, yet his ideas often were not taken seriously by other physicists. Perhaps this was because of his jocular irreverence toward his work, along with his penchant for irritating practical jokes. Or perhaps, unfortunately, it was because of his well-known alcoholism. Another problem for Alpher and Herman was that soon after making their prediction, they both withdrew from academic physics. Alpher took a job with the General Electric research labs in 1955, and Herman went to General Motors the following year. In those days, both labs were extremely highly regarded and supported fundamental research as well as applied work. Herman moved to the University of Texas in 1979, and Alpher moved to Union College in 1986.

Their departure from academia coincided almost exactly with the period in which cosmology developed from an interesting speculation into a respected science. Without cadres of graduate students and postdoctoral fellows to help develop their ideas further and then go out and proselytize for their mentors, Alpher and Herman had virtually no allies in the academic world. Indeed, neither was ever even invited to speak at a single gathering of astronomers or cosmologists. Alpher and Herman still blame themselves, in part, for their failure to include a graph of the spectrum of the blackbody with their prediction. Such a graph, they believe, would have made their calculation easier to comprehend and encouraged theorists and experimentalists alike to investigate the cosmic background radiation further.[9]

More important may have been the intellectual climate of the day. The idea of a primordial explosion leading to an expanding, evolving cosmos, then being championed mainly by Gamow, was simply not established as serious science. Most astronomers believed Gamow's creation scenario to have too many scientific drawbacks. For one thing, the theory did not explain the creation of the elements other than helium; nor did it explain the heavier elements. Another problem was the time frame. Hubble's expansion parameter (today usually called the Hubble constant) as it was being measured in the 1940s and early 1950s suggested that the universe was only about two billion years old. Yet geophysicists already had shown by the 1950s that rocks on the surface of the earth were far older than that. The inconsistency logically troubled most astronomers.

And an opposing theory had surfaced—one that at the time was at least equally attractive. In what may be an apocryphal story that has become part of cosmology lore, Fred Hoyle and two other young scientists in Cambridge, England, Herman Bondi and Thomas Gold, had seen a film called *Dead of Night* with a story line that ended exactly where it had begun. This supposedly gave Gold an idea: Might not the universe be scripted the same way—without beginning or end? From this simple concept, Hoyle, Bondi, and Gold developed what they called the steady-state theory.

This pictured a universe infinite in both past and future time as well as in space. The galaxies that Hubble had discovered flying away from us were continuously being replaced by the creation of new matter in the steady-state theory. New stars and galaxies would eventually form from this matter, and they too would rush away from one another. Such a universe, argued Hoyle, Bondi, and Gold, would always look approximately the same to any observer at any time during the universe's history.[10] Hoyle was (and still is) a forceful opponent of the Big Bang theory, whose name he coined on a British Broadcasting Corporation program in 1950.* The next year Pope Pius XII, upon learning of the Big Bang theory, said, "True science to an ever increasing degree discovers God as though God were waiting behind each door opened by science."

"What kind of scientific theory is this that was conceived by a priest and endorsed by the pope?" Hoyle wondered. Sarcastic, righteous, and possessed of a grating accent that was a peculiar blend of Midlands working class and Oxbridge, he was bright and iconoclastic.

---

*An article in *Scientific American* (March 1995) asserts that Hoyle did not intend to disparage the theory, but merely to make it more vivid. In 1994, *Sky and Telescope* magazine held a contest to rename the Big Bang theory. The magazine received thousands of entries, but was unable to pick a clear winner. Hoyle was not surprised: "Words are like harpoons. Once they go in, they are very hard to pull out."

He had learned the multiplication tables before he was three, and had worked his way through Cambridge University. He argued over and over against the Big Bang at scientific meetings and seminars with a voice dripping incredulity at the belief of other physicists in a theory as monstrously flawed as the Big Bang.

He also liked to point out that his theory was far sounder philosophically: After all, it required only that matter be continuously created to form new galaxies, not that all matter for all time be created in a single instant. Where did Hoyle's matter come from? He maintained that it would pop into existence in the void of space in the form of hydrogen atoms that coagulate into stars and galaxies. According to him, this was entirely possible according to the laws of quantum mechanics, which was coming of age in the 1950s and which Hoyle was willing to push further than most other astrophysicsts.

It did not help adherents of the Big Bang that Gamow was its most vocal supporter. Or that Einstein, now living out his remaining years in Princeton as the world's most famous scientist, was still philosophically more comfortable with a static universe. Or, most important, that Alpher and Herman's prediction of the cosmic background radiation, which could not plausibly be accounted for in steady-state cosmology, had been all but forgotten during the 1950s. With problems on both sides, neither was a clear winner. This was how matters stood until the early spring of 1965, cosmology stalemated.

A DAY WITHOUT YESTERDAY

Had Arno Penzias and Robert Wilson known in 1964 of the prediction of Alpher and Herman sixteen years earlier, the two Bell scientists would have been spared a year's work trying to uncover the source of the noise in their horn antenna. Had Dicke been aware of the prediction, he could have begun work on his own antenna years earlier without having to wait for Jim Peebles to do the theoretical calculations from scratch. But on the warm March Friday that Dicke along with his two young colleagues, Dave Wilkinson and Peter Roll, drove over to Crawford Hill, neither group was aware of the earlier work. Had they been, the events of that day would probably not have taken place.

The meeting was held within a few days of Penzias's call to Dicke, probably Friday, March 26, 1965, although nobody seems to remember exactly. The three Princeton scientists joined the Bell researchers in Penzias's office. Peebles had remained home because, as he explained later, he "wouldn't have had the slightest idea what they were talking about as far as that horn was concerned."[11] Dicke and his colleagues listened as Penzias and Wilson explained how the antenna worked and what kind of excess noise they had encountered. This noise, they said, corresponded to about 3° Kelvin. This almost exactly matched Peebles's theoretical calculation. Penzias and Wilson finished their explanation.

"They've got it," Dicke said, turning to Wilkinson and Roll.

Penzias recalled thinking that the Princeton group probably thought he and Wilson were "just a couple of telephone scientists." Roll was the most interested in their work on the antenna, while Wilkinson seemed a bit arrogant, Penzias remembered.[12] The five scientists decided to go outside and see the horn antenna. After a cursory look lasting all of "about two minutes," Penzias recalled, the group returned to Penzias's office to decide what step to take next.

At this point in the story, the recollections of the participants diverge significantly. Wilkinson recalled that Dicke suggested they all write a paper together. But Penzias and Wilson were not comfortable with this approach. Wilkinson remembered thinking that Penzias and Wilson were extremely cautious about Dicke's cosmological explanation for the noise in their antenna and were quite concerned about the possible negative reaction of their employer.

"They seemed to think that the cosmic background radiation was a pretty wacky idea," Wilkinson recalled. "In the end, we wrote two papers. And the rest is history."

According to Penzias, however, it was he who suggested that the Bell and Princeton scientists write a joint paper. "Dicke rejected it on the spot," Penzias said. He and Wilson were "very disappointed" in the final decision to write two distinctly separate papers on the discovery of the cosmic background radiation.

The two papers finally agreed upon were written quickly, and published within a few months in the same issue of the *Astrophysical Journal*. In their paper Penzias and Wilson described their serendipitous discovery of the cosmic background radiation. Dicke, Peebles, Roll, and Wilkinson explained in their paper the cosmological implications of the Bell scientists' discovery.[13] A single sentence in the Dicke paper noted the work of Gamow, Alpher, and Herman on primordial nucleosynthesis during the

© Eli Dwek

I wish He wouldn't keep that
darn thermostat at 3 K!

1940s; there was no mention at all of Alpher and Herman's accurate prediction of the temperature of the cosmic background radiation.

Not often in science do competing hypotheses rise or fall as the result
of only one experimental finding or discovery. This is something that
should occur more frequently, according to the ideas of Sir Karl Popper,
an Austrian-born philosopher of science who died in 1994. Popper
argued that science *should* be a process in which various hypotheses are
created to predict natural phenomena that then can be tested or observed.
If the prediction fails, then the hypothesis should be abandoned.

Thomas S. Kuhn, a historian of science at Princeton at the time
Dicke and his colleagues met with Penzias and Wilson, argued effectively that Popper's scenario rarely happened in the real world of scientific enterprise. In fact, Kuhn maintained, it was an idealized way of
looking at scientific progress. According to Kuhn, who coined the term
"scientific paradigm" in his 1962 book *The Structure of Scientific Revolutions,* acceptance of a new hypothesis among scientists usually occurs
only as adherents of an older, no longer valid hypothesis either die or
lose influence within a scientific community.

---

*Eli Dwek is a theoretical astrophysicist and a member of the COBE science team.

COLLEGE OF THE SEQUOIAS
LIBRARY

Neither Kuhn's nor Popper's concept of how science progresses—or, in their opinion, *should* progress—applies fully to the case of the competing steady-state and Big Bang theories. This is because neither of these hypotheses was entrenched within the astronomy community. Astronomers were about equally divided in their support of each one. Yet, as events unfolded, Popper's idealized principle seemed to predict the outcome more than Kuhn's more cynical concept.

With neither theory fully acceptable to astronomers before the discovery of the cosmic background radiation, the scientific credibility of the competing hypotheses was the determining factor—not the prevailing cultural climate among astronomers. In the light of the new discovery, the Big Bang theory was the clear winner for the simple reason that the steady-state model did not predict and could not reasonably account for the presence of the cosmic background radiation. On the other side, the Big Bang theory not only predicted the background radiation but required it.

Opinion among astrophysicists shifted almost overnight. Within just a year or two virtually all publication of papers espousing steady-state ideas in scientific journals ceased, except for a series from Hoyle and his colleagues attempting to solve their theory's problems related to the cosmic background radiation.

✦ 5 ✦

The Ballooning Universe

In science the credit goes to the man who convinces the world, not to the man to whom the idea first occurs.

—Sir William Osler

D AVE WILKINSON AND I first met in 1968. Swarthmore College had the unusual practice of bringing in outside examiners for seniors in its honors program. Dave was one of my examiners in physics. He had developed a fine reputation at Princeton as a young physicist with a future. He made a big impression on me, although at the time I had little idea that he had been involved in the discovery of the cosmic microwave background radiation. I performed well on the oral examination, earning highest honors and election to Phi Beta Kappa, then selected Princeton for graduate school because its physics and astronomy departments were among the finest in the nation.

But my friends at Princeton complained about the distance they had to travel to visit their girlfriends or to meet girls, and discouraged me from enrolling. In the meantime, Ted Chang, a friend I had met on one of my summer programs during high school and who was now at Berkeley, sent me an application for a summer job at the Lawrence Berkeley Laboratory. I took the job and enjoyed California so much that I decided to stay on at Berkeley. As the holder of a National Science Foundation graduate fellowship, I had movable funds: It was a reasonably simple matter to make the switch from Princeton to Berkeley in time for the fall term.

Initially, I had hoped to become another Richard Feynman, the winner of the 1965 Nobel Prize for helping develop quantum electrodynamics. I realized, though, that the job was already taken; moreover, my advisers were warning me that if I wanted to be an elementary particle theorist, I had better plan to be independently wealthy since there were virtually no positions available anywhere. Like many of my contemporaries in those heady days of the late 1960s, I wanted to be socially useful and thought about going into fusion reactor physics. This seemed to provide the most potential for helping the world in that fusion conceivably might eventually provide an unlimited supply of energy. I also thought about going on to law school. But as I looked over a catalog from the Berkeley law school, I instantly lost interest.

Meanwhile, the war in Vietnam had come to Berkeley. I tried to avoid the conflicts that were occurring more and more frequently around campus. As far as I was concerned, both sides seemed hopelessly naive and self-serving. And they were getting in the way of my scientific research. The campus hospital was teargassed (accidentally, of course) by helicopters sent in by Governor Ronald Reagan. Matters reached such a terrible state that both sides backed down. About the same time, I was examined for the military draft, but was too nearsighted to be selected.

Professionally, my interest had shifted from particle physics. During my second year at Berkeley, I ran into Charles Townes, a Nobel laureate for his work on masers. He was working with his postdoctoral fellow, Mike Werner, and a young faculty member named Paul Richards. Their work fascinated me. After receiving an invitation, I decided to join their group. Paul agreed to become my faculty adviser. Little did I know that this decision indirectly would set me on the course I was to follow for the next twenty years.

Townes and Werner were just preparing to take new measurements on the cosmic microwave background, which by then was a hot research topic. Another group had made a measurement of the blackbody spectrum showing an unexpected spectrum line, a wavelength where the radiation was especially intense. We set out to see if we could do a better job. I initially thought this would be a fairly simple thing to do. Unfortunately, the Fabry-Perot interferometer we incorporated into our system was subject to numerous errors. The worst of these was that its optical elements were metallic screens that had to be stretched tautly and securely across small frames. The screens were similar to the kind used in electric shavers but thinner and extremely fragile. Our second, and bigger, problem was that the atmosphere near Berkeley was opaque at the wavelength we wanted to study. We looked around for a more suitable site for our instrument.

We selected White Mountain, a 14,250-foot peak in the range east of the Sierras near Bishop, California. Its summit was all but inaccessible in winter except by helicopter. The ride up was terrifying as the helicopter, heavily laden with us and our equipment, could manage to take off only by nosediving off a small hill. The experimental station at about 12,000 feet was usually quite cold, and it often snowed there even in the summer. We worked in heavy parkas and boots, shivering as we tried to ward off the cold and the wet. We did manage to make a new measurement of the cosmic background's spectrum. But the paper we published as a result of the experiment made little impression on anybody, since our answer was the one everybody already expected: The earlier group's puzzling measurement was wrong.

As we continued working on the cosmic background, we learned that this was par for the course. Our work invariably seemed to be propelled forward by a sequence of wrong or implausible results. Our next experiment was more ambitious. A few years earlier, a combined team from the Naval Research Lab near Washington, D.C., and Cornell University had sent a detector aloft on a rocket with the intention of measuring the shorter wavelengths of the cosmic background radiation at high altitude. They reported that their instrument had found a total cosmic background energy some fifty times greater than that which almost everybody in the background community believed it should be.

We felt certain we could make a more accurate measurement. Of course, this meant that we would be doing a lot of work just to prove wrong an experiment that we and everybody else already believed was wrong. Paul Richards had heard about a polarizing Michelson interferometer that recently had been invented in England by D. H. Martin and E. Puplett.[1] It was a device quite unlike the Fabry-Perot interferometer we had already built. The Fabry-Perot incorporated a filter that would respond to only one wavelength at a time. We could measure a spectrum by tuning the wavelength of the filter. The polarizing Michelson interferometer encoded the incoming spectrum, producing an interferogram. This would not look like the spectrum we were after, but there was a simple mathematical relationship called a Fourier transformation that could decode the interferogram. We thought it might be far superior to the Fabry-Perot instrument, but we were a little suspicious because the device that had measured the false spectrum line in the cosmic background also had been a Michelson interferometer.

Paul's idea was to send one up on a balloon payload, high above the annoyances of breathable air in order to get clean data. Our British competitors, J. E. Beckman and E. I. Robson, had the same idea. Paul agreed with them that we would not exchange any notes. Whoever got the answer first would then know they had won fair and square. And, of course, there would be less chance that both groups would fall into the same error together. Paul designed our instrument from the Martin and Puplett concept. David Woody, another grad student in our group, and I built it with help from the physics department shop. Norm Nishioka, a third grad student who had joined us by then (and is now a medical doctor), worked with us.

We wanted to do a short wavelength measurement that could tell us whether the cosmic background radiation has a precise blackbody spectrum or only an approximation of one. None of the ground-based experiments at longer wavelengths could make such a determination, since they produced what essentially was a featureless parabola. There were a number of other cosmic sources such as dust in distant galaxies that possibly could produce the same parabolic effect. If we could make a good measurement at wavelengths less than about 2 millimeters, we would know the curve was not simply a parabola, but could really be the cosmic blackbody radiation as we expected.

HASTY SCIENCE

We launched our instrument from the National Scientific Balloon Base in Palestine, Texas. A typical launch costs approximately $100,000 in the United States and today is usually funded by NASA. In the 1970s the balloon base was operated by the National Center for Atmospheric Research. The crew, based permanently in Palestine, planned to launch our balloon at dusk when the winds were relatively calm. High-altitude scientific ballooning is a truly astonishing sight. The balloons are enormous, approximately 100 yards in diameter—as big as a football field.

Each balloon consists of thousands of pounds of polyethylene plastic less than a thousandth of an inch thick, about the thickness of a dry-cleaning bag. The material is extremely fragile and tears easily if it catches on a twig or bit of straw on the ground. Yet it is capable of lifting thousands of pounds into the upper atmosphere. Only a few manufacturers in the world have the expertise to assemble these giant bags, then pack them into crates so they can be opened at the launch site in one piece.

We watched nervously as our balloon was prepared for launch. It was laid out on the ground and held down by a huge roller. The crew began pumping a small bubble of helium gas into one end from a truck loaded with large cylinders. The payload with our detector waited hundreds of yards away, affixed to the bottom of the nearly empty bag by cables and a parachute. At last the moment arrived. The great roller released the balloon and it began rising from the ground like a volcanic island growing out of the sea. Meanwhile, a large modified earth-moving machine named Tiny Tim raced across the launchpad, our payload suspended from its jaws until the balloon itself took over.

At that instant the mammoth gas bag with our precious payload suspended beneath it began rising slowly and majestically into the evening sky.

The ascending balloon soon vanished. Back in the control room we ascertained that it was traveling eastward, which, owing to the prevailing winds, was normal. Unfortunately, we were rookies in the balloon business and had yet to appreciate the wisdom of Murphy's law. When our interferometer had worked well on the ground back in Berkeley, we had became impatient to fly the beast without testing everything we could possibly test, too quick to load our detector into the back of a truck and head off to Palestine. As we drove into west Texas, we paid little attention to what some people might have considered an ominous event: A large black bull rose from the side of the highway and walked directly into our path. We stopped the truck short of hitting him by less than a yard.

As our balloon flew east out of Texas, we determined that a small motor that moved the mirrors had stopped working. Also, the amplifier that increased the signals from the detector had begun making its own oscillations. The amplifier was now transmitting a signal that had absolutely nothing to do with the blackbody spectrum. We desperately tried to fix it with radio commands, but it was hopeless. Nothing worked.

We were exhausted from our all-night vigil. By the time the balloon was over Alabama, it was time to bring down the payload. A radio signal was sent, and our wonderful instrument that had detected nothing of interest to us was severed from the balloon and floated to the ground. The flight had been perfect. The balloon had flown as planned, and the detector arrived safely back on the ground, protected from impact by crushable cardboard pads. The instrument itself turned out to be the culprit.

When the instrument failed, my heart sank. There goes my doctoral dissertation, I thought. Fortunately, Paul took a benevolent view. He and the university agreed to let me go ahead and write my thesis on our successful mountaintop experiment and the design of our balloon experiment, even though it had not worked as planned. In the end I received my Ph.D. essentially for the construction of the instrument rather than for any data.[2]

After I left the group, Dave, Paul, and Norm checked out our apparatus properly in a plywood and Styrofoam dry ice chamber. They made appropriate changes in the detector, then flew it successfully. They found a small excess near the peak of the cosmic background spectrum, which we were obliged to report but which we never really believed. The signal was somewhat brighter than it should have been around 1-millimeter wavelength, and we were worried that there could be some fault in the design of the instrument. We never learned why. Years later Jeff Peterson, another of Paul Richards's students, told us we had underestimated the radiation emitted by the warm parts of our apparatus. I now think he was right. And the results were still closer to a perfect blackbody than anybody else had obtained. We took the sensible point of view that we had tested the Big Bang origin of the radiation and found it was just what it was supposed to be.

Sending these big, high-altitude balloons aloft with one kind of instrument or another on board was the way a researcher made his name in the early days of cosmic background exploration. But pitfalls awaited the careless researcher, as I had learned painfully. At least two British groups were working on similar kinds of measurements with balloons and spectrometers during the early 1970s. One group, led by John Beckman and Ian Robson, worked around the clock to get their balloon up before anybody else, then rushed to get their results into *Nature,* a British journal noted for its ability to get exciting results into print quickly.[3]

Unfortunately, in their haste they inadvertently deleted from their text an important section pertaining to the effects of atmospheric emissions on the cosmic background. In their published report, there was no sign of these lines and no discussion of what had happened to them. Their spectrum of the cosmic microwave background radiation seemed dubious at best. Paul Richards, who was well respected throughout the cosmic background community and could get away with it, later chided them regularly about the importance of careful work and careful reporting.

Rainer Weiss, an MIT physicist whom I would come to know quite well a few years later, had gotten into the spectrum business long before the Berkeley group. Ray had returned to MIT from Princeton in 1965, just when the cosmic background radiation work was getting exciting there. At the time he was working on the experimental consequences of a scalar gravitational field and looking into measurements of slow changes of the gravitational constant. At MIT he was asked to teach general relativity, one of his interests, so he went into it in detail and started thinking about cosmology.

One of the students in the course was Dirk Muehlner, who was looking for something more exciting than far infrared spectroscopy of solids, which he was then doing. When they heard about the discovery of the cosmic background radiation, Ray and Dirk realized that, with Dirk's experimental expertise, they might be able to see the Planck spectrum peak around the 2-millimeter wavelength. In their minds, that would clinch the radiation's cosmological origin. Bernie Burke, who had helped get Penzias and Dicke together, now had begun his own low-frequency measurement program. He encouraged Ray and Dirk to try a measurement at higher frequencies. Ray talked over the idea with his colleagues in the radio astronomy group who knew something about molecular spectroscopy of the atmosphere. It soon became clear to him that the experiment could not be done from the ground.

Ray borrowed some ballooning equipment from the radio astronomy group (which was never returned, since it perished in free fall). By its nature, ballooning is not a certain thing. There is a nagging worry that residual water vapor at high altitudes will ruin the measurements, as already indicated by rocket-borne instruments. Ray did not believe it, and went to visit some of the individuals who had launched the rockets. After talking with them, he concluded that a rocket probably carried up a little cloud of vapor as it ascended. The data had to be wrong, he concluded.

Ray and Dirk decided they would have to use a cryogenic system immersed in liquid helium to keep the optics of their instrument cold enough to prevent the instrument itself from radiating detectable energy. They came up with a rather strange design in which the optics were encased within a sealed can filled with liquid helium that would maintain thermal conductivity. The can itself was placed within a big, open-mouthed dewar resembling a large thermos bottle without a top. Liquid helium within the dewar would be free to evaporate, keeping the entire instrument package cool. They also incorporated broad band detectors, wavelength filters they built themselves, and a mechanical chopper disk to change the intensity of the incoming light in a periodic way. Their biggest worry was how to keep the high-altitude atmosphere from condensing onto the apparatus, making it resemble a Russian winter scene. They also worried about possible effects of the balloon itself, so they wound a 2,000-foot rope on a reel in order to lower the payload after launch.

Their first flight went beautifully. The apparatus confirmed that the cosmic background had the expected Planck blackbody form at low frequencies, but showed an excess around the peak that was later shown to be incorrect. To account for atmospheric emissions, they also measured the signals as the apparatus was tilted to account for varying conditions in the air. The result, although not perfect, was tantalizing. They rushed to launch a second balloon three months later. The instrument worked well. But the payload somehow received a false command, plummeting it to Earth without a parachute and leaving them without data from the flight.

They published the results in the *Physical Review* in 1973, showing clearly that the spectrum had the expected blackbody peak and curvature. The only residual doubt involved possible atmospheric effects and the size of the error bars (uncertainties), which ideally could have been smaller. Still, Dirk and Ray had it first.[4] Their positive result was among the pieces of news coming from the cosmic background radiation community that propelled our Berkeley group up White Mountain.

All the researchers spent a good part of their time simply making sure their equipment worked properly. So, of course, it was necessary to stay abreast of the steady stream of refinements. Weiss and Muehlner, for instance, found that a plastic window used to keep air from condensing on the surfaces cooled by liquid helium was not necessary. They discovered this quite by accident when a small lead weight used for ballast fell through the plastic, which was thin as a laundry wrapper, and broke it. There was virtually no effect on the measured brightness of the cosmic background. After that, all the balloon experimenters began designing a plastic window they could open or close at will.

During these years a small group led by Herb Gush at the University of British Columbia in Vancouver had started doing a far more difficult experiment with a rocket. Although it was a grand idea to rocket a payload far above the atmosphere, its implementation was unusually difficult. For one thing, the size of the payload was limited, only about 1 foot by 2 feet, depending on the kind of rocket. The rocket size itself, of course, was directly dependent on how much money was available. Moreover, the flight would last only a few minutes. And the rocket could, of course, explode. Or, more likely, the parachute carrying the instruments would fail and the payload the researchers had spent months building would be smashed to smithereens upon impact. Sometimes even the weight of the parachute was too much for the rocket and none could be used.

Another problem for the rocket experiments was zero gravity: What would happen to all that liquid helium sloshing around when there was nothing to hold it down in one place? Gush's group, which had begun launching the rockets when I was at Berkeley, continued their work for nearly twenty years. Most of their flights failed for reasons incidental to their instrumentation, such as the rocket blowing up on the launchpad. Gush was a great unsung hero in the story. Eventually he and his student Ed Wishnow and postdoc Mark Halpern (who had been Ray Weiss's student) obtained some very fine data. Unfortunately, their results came in late, and Gush and his group never received much recognition for their work.

During the period I was working on the balloon experiment, a review committee from NASA came to visit the Space Science Laboratory at Berkeley. The lab had funded part of the project and even lent us one of its best engineers, Henry Primbsch, to help with construction. I gave a short presentation to the committee about the balloon flight we were planning in order to measure the cosmic background radiation. One committee member asked whether such measurements could be taken from a satellite. Nothing would prevent it, I said. If so, he wondered, why was it not being done and why was I not doing it?

I had yet to earn my doctorate. My answer to the question was tantamount to, "Who, me?"

THE LOST PENNY

During the years I was finishing up my thesis at Berkeley, the cosmic background radiation became a major fixture in cosmological observation. In fact, the cosmic background had opened a floodgate of scientific papers almost from the day its discovery was announced in 1965. One scientist who had not been immediately convinced by the discovery was Robert Wilson.

"My first notion that this was serious stuff occurred one day that May when my father was visiting from Texas," recalled Wilson, who had studied under Hoyle and was still philosophically disposed to a static universe.[5] "He went out and bought the *New York Times*. There on the front page was a story about our discovery and its significance."[6]

Walter Sullivan, a *Times* science writer, had heard about the detection of the cosmic microwave background from an editor at the *Astrophysical Journal*. He called Penzias for the details. Neither Penzias nor Wilson had thought much would come of Sullivan's call. In his article, Sullivan correctly explained the importance of Penzias and Wilson's discovery of isotropic radiation at approximately 3° Kelvin with a wavelength of just over 7 centimeters: The fact that the radiation was isotropic meant it was the same across the sky. This was precisely what those astronomers who supported the Big Bang hypothesis had expected.

Radiation that was a remnant of the Big Bang, which had occurred everywhere in the universe at once, *should* have been of equal intensity everywhere. Of course, a measurement at only one frequency could not prove conclusively that the radiation was blackbody radiation from the Big Bang (a finding that would not occur for twenty-five years). The discovery by the Bell Labs scientists soon was confirmed by the Princeton team at a shorter wavelength.

The blackbody nature of the background radiation was something Big Bang astronomers also expected. As we've already seen, this kind of radiation occurs whenever particles collide in a thermal equilibrium of the kind that might occur inside a black box. The Big Bang should have produced extremely intense particle collisions. Most cosmologists—not including Fred Hoyle, of course, and the others who still believed in a steady-state universe—were now believers: A huge amount of blackbody radiation had been produced during the earliest moments of the universe and was still present in a sufficient quantity to detect (the test of its blackbody spectrum being, of course, one of the most important reasons for flying the COBE).

Robert Wilson met George Gamow for the first and only time in December 1965 at a hotel in New York City during the annual Texas Symposium on Relativistic Astrophysics. By then the discovery of the background radiation was already well established not only in astrophysical circles but in the public mind as well. Even the reluctant Wilson had started becoming a believer. Gamow seemed somewhat upset over the fact that no credit had been given him for the early theoretical prediction of the cosmic background temperature, Wilson recalled.

"If I lose a penny and someone finds it in the same place where I lost it, then I still know it's my penny even if I can't prove it," Gamow told Wilson.

Wilson was dumbfounded. This was the first time he learned that anybody other than Dicke and Peebles had worked out a theoretical scheme that would account for their famous accidental discovery with the horn antenna. Whether Gamow at the time sought scientific credit for himself or his colleagues Alpher and Herman is not clear. Gamow himself had not published a word about the background radiation before Alpher and Herman's 1948 paper, contrary to what has been written in many popular accounts, including one by Peebles and Wilkinson published in 1983.[7]

A 1948 paper in *Nature* by Gamow, widely cited as the first theoretical prediction of a 10°-Kelvin background temperature, was concerned with primordial nucleosynthesis and early galaxy formation. It made no mention of background radiation.[8] In any event, by early 1967 Gamow and his colleagues, Alpher and Herman, had become "very perturbed by how our early work continued to be ignored."[9] Gamow suggested to Alpher and Herman that they try to set the historical record straight. They made the attempt in a lengthy article the three wrote the same year.

Their choice of journals, *Proceedings of the National Academy of Sciences,* was unfortunate because it is a publication for *all* scientists and not for one specific discipline, which is what really interests any scientist worth his salt.[10] The piece documented Alpher and Herman's theoretical work on the cosmic background radiation in the 1940s as well as their joint efforts with Gamow on galaxy and star formation. But almost no astronomers or physicists read it. Gamow lived out his life in relative obscurity at the University of Colorado in Boulder. He died in 1968, still believing he had not received due credit for his early work on the origin of the universe.

By the mid-1970s Big Bang cosmology was almost universally established in the minds of astrophysicists and theorists. In the summer of 1978, Penzias read an article in *Omni* magazine stating that he would be awarded the Nobel Prize. "I laughed out loud," Penzias recalled.[11] The magazine, however, proved to be a good fortune-teller. That autumn Penzias and Wilson were awarded the prize for their unplanned but carefully worked discovery of the cosmic background radiation. Shortly before Penzias was to depart for Stockholm to receive the prize, he invited Ralph Alpher to visit him at his home in New Jersey. Penzias wanted Alpher to bring him up to speed on contemporary cosmology before delivering his Nobel acceptance speech. Alpher recalled the visit bitterly: "There was never any invitation to visit Bell, to see the radio telescope, to meet Wilson, nothing, no friendly overtures at all."

According to Alpher, Penzias seemed to regard Herman and him as part of the "folklore of physics," as if "being so dubbed should be sufficient recognition of our contribution."[12] Shortly after visiting Penzias, Alpher suffered a heart attack, which he has suggested may have resulted from the stress of the visit. There was more than enough unhappiness to go around over the way credit for the discovery of the cosmic background radiation has been dispensed over the years. The main culprits in the minds of Alpher and Herman apparently were members of Dicke's Princeton group who, Alpher and Herman felt, refused to give them due recognition even after becoming aware of their early work.

"Jim Peebles knew of our work, unless he was incredibly obtuse," said Alpher. "Peebles got from us two reviews of his paper on the background radiation well in advance of the Dicke-Peebles-Roll-Wilkinson 1965 publication." This was a reference to the Peebles paper that had been rejected by *Physical Review* and which, unknown to Peebles, had been reviewed for the publication's editor by Alpher and Herman.

"Some psychologist could have a field day with the fact that on two occasions [I] ran into Peebles at meetings: He professed no name recognition," Alpher said.[13] As for Dicke, said Alpher, his "excuse to us in correspondence that some of his work in other areas had been overlooked and why, therefore, were we upset, just does not wash." He believed that Dicke's attitude was, "If it wasn't invented here, it wasn't invented."[14]

Nor was Penzias particularly happy with Dicke. Penzias believed that Dicke and his Princeton group had not been eager to credit him and Wilson for their role in discovering the cosmic background radiation. An article Peebles and Wilkinson subsequently wrote for *Scientific American* magazine on the discovery of the cosmic background seemed to bolster Penzias's point. Although the work of Gamow, Alpher, and Herman was noted, Penzias and Wilson were barely mentioned.[15]

Peebles has stated repeatedly that he and Dicke simply were not aware of the earlier work of Alpher and Herman. Once he and Dicke learned of it, Peebles said, they began citing the earlier prediction of the cosmic background temperature. In a 1966 paper in the *Astrophysical Journal,* Peebles did cite as a reference a 1953 paper written by Alpher and Herman along with another physicist named James Follin. But he did not mention their work in the body of the text. This paper was a revision of the 1965 article that had been rejected by *Physical Review* for poor scholarship. Peebles did not learn until late 1994, moreover, that Alpher and Herman had reviewed the rejected paper nearly thirty years earlier. Peebles had thought all along that George Gamow had been the reviewer of the paper.[16]

"Everything considered, I think Bob and Ralph have been given the credit they deserve," Peebles said. He believed that Alpher and Herman undoubtedly had been hurt by their departure from academia for industry during the 1950s and by their affiliation with Gamow, whose reputation within the cosmology community was less than pristine.

Fortunately or unfortunately, depending on one's perspective, Sir William Osler, the renowned Canadian physician and medical historian, was right all along: Convincing others is more important than being first.

✳ 6 ✳

Tuolumne Nights

The Arbiter of others' fate
A Suppliant for his own!

> —GEORGE GORDON BYRON, LORD BYRON

I
N AN EXTRAORDINARY book called *Betrayers of the Truth: Fraud and Deceit in the Halls of Science,* William Broad and Nicholas Wade, reporters at the *New York Times,* maintain that deceit, fraud, and the unethical taking of credit for oneself occur every bit as often in science as in other endeavors. In what could be considered a view from the dark side, the authors allege that some of history's greatest scientists—Galileo, Newton, and Mendel, to name a few—verged on committing fraud. Today scientists commonly cheat, steal credit, and fudge their data in order to conceal their mediocrity, the authors contend: "By studying science through its pathology rather than through some preconceived criterion, it is easier to see the process as it is, as distinct from how it ought to be."[1]

I have little sympathy for such a bleak view of science. Yet I realize that the invisible mechanism by which credit for a new theory or discovery is dispensed within a scientific community—where credit is after all the coin of the realm—grinds its gears mysteriously. As we have seen and will see again, how credit for scientific work is given often differs considerably from how one might think it *should* be given. Consider some of the other possibilities had the story of the discovery of cosmic background radiation taken a different turn here or there. If, for instance, the Princeton and Bell Labs groups had agreed to write a joint paper, then Dicke very likely would have shared in the Nobel Prize.

"We would have been happy to have been authors number five and six on Dicke's paper," Penzias recalled.[2] What would have happened had either of the 1965 groups known of the Gamow-Alpher-Herman theoretical work? Perhaps one of the three would have been included in the Nobel. Or, Wilson believed, had Gamow been alive in 1978, he might have shared in the prize (although Gamow had not actually participated in the published prediction of the cosmic background radiation). Since the Swedish Academy limited the Nobel Prize in any one category to no more than three recipients, Wilson probably would have been left out had Gamow and Dicke been included—"a result I wouldn't have much liked," Wilson conceded. He thought it was likely that none of the theorists was included in the prize because of the dispute lingering then for over a decade after the cosmic background discovery—and still lingering today—over how credit for the various theoretical predictions of the background temperature should have been partitioned.

There were still more unfulfilled possibilities of the story of the cosmic background radiation discovery. In 1964 two Soviet scientists, A. G. Doroshkevich and I. D. Novikov, published an article in Russian, later translated into English, suggesting that blackbody cosmic background radiation could probably be detected with a microwave receiver like the one Penzias and Wilson were then working with at Bell Labs.[3] Doroshkevich and Novikov had come to this conclusion by putting together the earlier theoretical work of Gamow, Alpher, and Herman with technical papers from Bell Labs.

Unfortunately they apparently misread one of the papers from Bell Labs, perhaps because of a poor translation. As a result, the Soviet scientists failed to recognize that Bell scientists working on the horn antenna prior to Penzias and Wilson had already discovered the cosmic background radiation but did not know it. If they had read the article correctly and completely, perhaps they would have shared a Nobel Prize themselves with one of the Bell researchers.

On still another front of unrealized potential, a Canadian scientist named Andrew McKellar reported in 1941 that he had discovered a method in the 1930s of measuring the rotational temperature of interstellar cyanogen molecules. These molecules consist of just two atoms, one carbon and one nitrogen, bound tightly together and rotating like a little dumbbell. When the rotational energy is quantized according to the laws of quantum mechanics, the hotter the cyanogen molecule the greater its rotational energy. These molecules could be observed in interstellar space because they absorbed light from distant stars at specific wavelengths. Since the wavelength depended on the rotation of the molecules, it is possible to measure the rotation of the cyanogen molecules by studying the spectra of stars.

Using such a technique, McKellar found an interstellar temperature of about 2.3° Kelvin, which approximated that predicted by Alpher and Herman seven years later.[4] Thinking the temperature he had measured simply came from a number of unobservable sources, most astronomers at the time thought little of McKellar's detection of an apparent background temperature. It is "of little significance," stated one of the elders of spectroscopy. The temperature generally was assumed to be due to the excitation of the molecules by interstellar electrons, for instance, and it was not yet known that all stars would give the same numbers.

After Penzias and Wilson found the cosmic microwave background, McKellar's old data were reinterpreted. More precise measurements confirmed the accuracy of his measurements. During the 1970s astronomers routinely relied on an updated version of the technique McKellar had developed nearly four decades earlier. If Alpher and Herman had known about McKellar's discovery in the late 1940s, they might have been able to put their theoretical work together with his observations, in which case all three might have shared the credit for the discovery of the cosmic background radiation and shared a Nobel Prize as well.

The most seductive alternative possibility in the story of the cosmic background discovery is to imagine what might have happened had Dicke learned of Alpher and Herman's work in the 1940s, or vice versa. In either case, Dicke, Alpher, Herman, or even Gamov would have recognized on the spot that the cosmic microwave background could be measured as Alpher and Herman had predicted it. Cosmology would have been born as an empirical science nearly two decades earlier than it was. And many subsequent events either would have become meaningless or would not have occurred at all. Dicke, Alpher, and Herman could have received all the credit, while Penzias and Wilson might have remained obscure Bell researchers.

Such was the tortuous path of scientific reality, however, that Dicke, Alpher, and Herman, although certainly the private heroes of the story, never received the full public recognition they and their supporters believed they deserved. Moreover, the discovery process for the cosmic background radiation was not really a "process" at all, for that word implies a series of deliberate steps. Yet by the end of 1965 cosmology, which until then had been a suspicious enterprise to many scientists, had become a true empirical science. The early universe was now a kind of laboratory for whatever kinds of theories and tools scientific researchers were clever enough to devise. This laboratory had been opened by Alpher, Herman, Gamow, Penzias, Wilson, Dicke, and others. Most important, it appeared to provide an answer to the big question: Did the universe have a beginning?

Today most scientists believe that the universe did begin in the Big Bang. Yet many, many questions remain. Was the Big Bang the true beginning of the universe? What do words such as *beginning* mean? Had the universe evolved from an earlier oscillation phase or something even more exotic, as theorists such as Robert Dicke have suggested? Will the universe continue to expand indefinitely? Or does it contain enough matter to bring it collapsing back on itself in a great gravitational day of reckoning? What is the role of black holes in an evolving universe?

Most intriguing of all: What happened to precipitate the Big Bang? Technically speaking, such a question is misleading. It seems to imply a possibility that the universe sprang into being out of nothing at a certain instant that astrophysicists describe as time $(t) = 0$, the Big Bang. All we really can know from extrapolating back in time is restricted to $t > 0$ (time greater than zero), so we could say nothing about a creation event itself if there had been one. If there was something leading up to such an event, then the evidence for it very likely would have been scrambled beyond recognition in the primal burst.

As the years passed and measurements improved, it became generally recognized that there was one big problem left unresolved by the events of 1965: The cosmic microwave background radiation seemed far too smooth to account for the universe we see now. The very uniformity that proved the cosmic origin of the microwave radiation was becoming too perfect. In other words, the background temperature seemed to be dispersed much too uniformly to have allowed for the formation of galaxies as we see them today billions of years after the Big Bang.

GUMSHOE SCIENCE

Astronomers from up and down the West Coast, and a few from the East Coast, gathered in Yosemite National Park in August 1972 to discuss the state of the cosmos. We met in small groups at Soda Springs in Tuolumne Meadows. The weather already had started changing by then, and toward the end of the session it started to snow. One morning it was so cold that we sat outside in our parkas to discuss what was new about the universe. Afternoons generally were free, and most of us went walking in the mountains. Some of the more skilled members of the group went rock climbing.

My lab partner, David Woody, was an accomplished climber. I was in awe of him and the others as they scrambled up cliffs that I would have been too worried about falling down to go out and try climbing myself. I console myself with the knowledge that I'm still alive and some of the more adventurous climbers are not. Perhaps other people sometimes think about scientists the way I thought about the rock climbers: simply beyond them. Of course, science has always seemed like a natural enterprise to me. And I suspect that scampering up a sheer rock surface without apparent effort is every bit as easy for a natural-born climber.

In the evenings we socialized over dinner, then gathered to hear a lecture. Pat Thaddeus, a senior research physicist at the Goddard Institute for Space Studies and an adjunct professor at Columbia University, gave one of these talks on interstellar molecules. Pat, who had received a masters degree in theoretical physics from Oxford University and a Ph.D from Columbia in 1960, had done some fine work on these interstellar molecules and their relationship to the cosmic background radiation. A superb storyteller and inspirational speaker, he was characteristically enthusiastic at Tuolumne. I thought to myself I would really enjoy working with him and learning something about radio astronomy.

Another night Maarten Schmidt, who recently had discovered quasars with the 200-inch Palomar telescope in southern California, gave us an evening course on constellations. This may seem strange, but most of the astronomers there lived in or near cities or towns and could not recognize anything in the night sky but the most major constellations. In cold, clear mountain air, it was virtually impossible to do even that. There were so many more stars visible than in the city that a familiar pattern simply disappeared among the myriads.

A year or so after the Yosemite meeting, Pat Thaddeus called me from New York and offered me a job as one of his postdoctoral fellows. It seemed like a wonderful opportunity to work with a great astronomer and a great man. I accepted on the spot. I told him I would have to wait for the completion of our balloon flight so I could finish my thesis. I assured him it would not take long. As it turned out, I barely made the deadline a year later, leaving Berkeley for the East Coast with my new Ph.D. at the end of January 1974. I moved into an apartment on West End Avenue in Manhattan and biked to work at the Goddard Institute for Space Studies, a NASA-run laboratory on 112th Street and Broadway. I had felt somewhat frustrated by our balloon work at Berkeley, and was interested at first in switching into another field where I could obtain data and write a research paper more quickly. Pat suggested I look into masers.

The name was an acronym for "microwave amplification by stimulated emission of radiation." A maser could be either a device in the lab or a natural mechanism in a star by which atoms are kept at abnormally high energy levels so that they will greatly amplify radio signals. A maser in a star is arranged in such a way that enormous amounts of energy are emitted from a very small volume of space at one particular frequency. Silicon monoxide masers had just been found in the red supergiant stars that were losing their stuffing at such a great rate that the entire star would disappear quickly in cosmic terms—just a hundred thousand years or so.

Charles Townes, whom I had worked with in Berkeley, first discovered the principle in the lab when he had been at Columbia a number of years earlier. He used multiple electrostatic plates to separate high-energy ammonia molecules from low-energy ones. Then he radiated the high-energy molecules with radio waves, which he found stimulated them into emitting similar radiation with a strengthened signal. (It subsequently was discovered that the very narrow frequency emitted by the ammonia maser made it one of the most accurate atomic clocks.) Townes won a Nobel Prize for his discovery and took out a number of patents on masers and lasers, which were related. The natural masers in space could not be patented, of course, and no one knew exactly how they worked.

Masers seemed like a good project for me. Most people, particularly those who are not scientists, probably think the greatest glory in science goes to the theorist with the idea that pulls everything together, as Einstein had done with his special and general theories of relativity. But I always have preferred the real stuff: hardware that can give us a new look at something nobody has ever seen before. Some theorists are so imaginative and so mathematically ingenious that they are able to create an explanation for virtually any set of facts, no matter how apparently inconsistent. In fact, it has been said that the mark of a good theorist is that very ability.

The difference between theorists and experimentalists or instrument builders like me often seemed a little like the differences among the archetypes we can read about in detective stories. Some detectives seem to have a clear picture from the very start of how the mystery will be solved, like Sherlock Holmes able to answer questions others had not yet been able to pose. Another kind of detective, a gumshoe like Sam Spade, relies on diligence and blind determination to uncover the evidence that will eventually crack the case. In cosmology just as in mystery stories, we celebrate the former and rely upon the latter to find out exactly what happened.

Too much reliance on sheer intuitive ability can cause trouble when investigating the natural world, I have always thought. Einstein himself got into real difficulty when his theory of general relativity seemed inconsistent with the observed world. He relied too much on his intuition and too little on observations about the real universe. This caused him to overlook entirely the possibility that the universe was expanding. His intuition had convinced him it was impossible, making him miss what would have been the most remarkable prediction in the entire history of science.

Most of the advances throughout the history of astronomy have been based on observations. These observations were then followed by theoretical synthesis. Einstein's relativity was the exception. In the case of Nicholas Copernicus, his marvelous heliocentric synthesis was based upon observations over many years that conflicted with the existing

Ptolemaic Earth-centered cosmology. Martin Harwit, an astronomer at Cornell University,* documents the progress of astronomy in a fascinating book called *Cosmic Discovery,* in which he classifies domains of discovery according to wavelength range and several other criteria.[5]

Harwit discovered that most important astronomical discoveries were made with instruments developed for other purposes or heavily dependent on technology developed for another use, frequently military in nature. The great Archimedes was employed by the tyrant of Syracuse to build war machines in the third century B.C., for example, and Leonardo da Vinci designed military machines. As we will see, this was certainly true of the entire U.S. space program. It would not have reached its prominence today without the stimulus of the pioneering and terrifying Nazi rocket experts, the cold war against the Soviet Union, or the ongoing requirements by the Pentagon for instruments using radar and infrared technology that could see through darkness and clouds.

My picture of good science is somewhat simpler. It begins with obtaining good, hard data with new and ever more ingenious instruments that will induce Mother Nature to yield her secrets to us. To help me uncover the secrets of the cosmic masers, I built a microwave receiver with the help of the engineering staff at Goddard, especially Sam Palmer and Dennis Mumma. A machinist named Irving Silverberg worked real magic in constructing the instrument's tiniest parts. A remarkable man, he had invented doubled-sided adhesive tape and was independently wealthy. He worked in the lab simply for the fun of it.

After it was built, I took the receiver to the MacDonald Observatory in Texas, where I had a great deal of trouble making it operate correctly. I succeeded in eking out a small amount of data, which I later incorporated in a relatively insignificant paper. I subsequently mounted the instrument on the 85-foot Naval Research Lab telescope on the shore of the Potomac River in Maryland. Again, I had little success. The receiver I had designed simply was not up to the task of detecting stellar masers. Little ever came of the few measurements I did manage to take.

As so often seems to be the case, when one door closes another one

---

* Harwit was later director of the Smithsonian Institution's Air and Space Museum in Washington, D.C.

opens. I had kept my chance meeting with the NASA committee in Berkeley in the back of my mind, along with their suggestion that I look into doing a cosmic background experiment aboard a satellite. The chance came much sooner than I expected. On July 15, 1974, when I had been in New York only about six months, NASA issued two calls, or "announcements of opportunity No. 6 and No. 7," soliciting proposals for scientific experiments to be placed in orbit on a satellite. The experiments would be launched by either a small Scout or medium-sized Delta rocket as part of NASA's Explorer program for scientific research in space. The announcements suggested that the experiments should be designed to study Earth, the Sun and the planets, the galaxies, or the most distant regions of the universe.[6] It was a virtually open invitation. Pat Thaddeus showed me the NASA announcements. "Why don't you see if you can come up with something?" he said.

The timing could not have been better for the study of the cosmic background radiation—or for my career. Of course, the only thing I knew much about as far as cosmology research was concerned was the blackbody spectrum of the cosmic microwave background, owing to the failed balloon experiment in Berkeley. At that very moment my Berkeley buddies were preparing it for its next flight. This experiment seemed to be begging for a ride into space, where it could get far above the sheltering sky.

This looked as though it could provide me with a means of contributing to our most fundamental knowledge and help answer a basic question about the universe: What were the dynamics of its creation? By 1974 the background radiation had become an essential part of observational cosmology, having passed from its discovery period of the late 1960s into a new analytic phase. The radiation was so uniform—or isotropic—in any direction it was measured and its spectrum so close to that of a blackbody that it seemed to satisfy almost beyond a doubt the theory that it was the modern remnant of a primal explosion.

Yet its pristine state cried out for more detail. The radiation had unique properties that I believed we could measure with precision instruments that might reveal to us the imprints left by early cosmic processes. After doing some rough calculations, I felt reasonably confident that I could design an instrument a thousand times more sensitive than the one we had tried unsuccessfully to launch in a balloon when I was at Berkeley.

I went to see Pat with my idea for the new device. He looked over my calculations. "I think this might work," he said, to my relief.

Each proposal submitted to NASA required the backing of a team of scientists. My first job would be to form such a team. Pat gave me the names of several of his colleagues working at NASA's Goddard Space Flight Center just outside Washington, D.C. He also told me about a number of other people around the country doing research on the cosmic background radiation. Pat himself was already an expert on the interstellar molecules of cyanogen and their use in measurement of the background radiation. I went back to my office and started making telephone calls.

THE MUSIC OF THE SPHERE

The two "announcements of opportunity" it issued in 1974 for science experiments to be launched into space were somewhat of a departure for NASA. Never before had the agency offered such a generous opportunity, even though scientific research had been a part of its charter since its founding shortly after the dawn of the space age. This had occurred when the Soviet Union launched a 184-pound ball called Traveler—Sputnik, in Russian—into orbit on October 4, 1957, a day the world changed forever. The satellite carried little more than a radio transmitter that broadcast rhythmic beeps from its protruding antennas as it journeyed around the earth every ninety minutes.

Anyone who was old enough in 1957 to remember Sputnik I's music of the sphere played over and over on the radio will recall a haunting sound carrying many images at once: an eerie sort of excitement of a new age to come, fear of the deepening cold war, the foreboding sense that the United States, still enjoying the relatively complacent 1950s, had fallen hopelessly behind the Soviets in science and military technology. The world, it was apparent, would never be the same.

The stage for the first artificial satellite had been set some fifty years earlier when, appropriately enough, a Russian schoolteacher named Konstantin Eduardovich Tsiolkovsky made a theoretical calculation of mass ratio indicating that space flight should be possible with rocket motors that did not require the intake of air like a jet engine.[7] Remarkably, Tsiolkovsky accomplished this after he had lost his hearing in childhood and then, not being permitted to attend the usual schools, educated himself.

Tsiolkovsky, who had read Jules Verne, was a visionary in his own right. In 1929 he designed what he called a "rocket train," or multistage rocket powered by liquid fuel that he thought would allow humans to "emerge from the bonds of the atmosphere." He proposed artificial manned satellites that could be used as refueling stations for interplanetary journeys or as laboratories where scientists could carry out experiments not possible on Earth. One of his designs even pictured a space station spinning around a central axis in order to create artificial gravity. He believed that even short periods of weightlessness might have terribly ill effects on space voyagers*

Tsiolkovsky died in 1935, not seeing his vision for an age of space exploration become reality. A liquid-fueled rocket, however, already had flown. It was launched in 1926 by Robert Hutchings Goddard, a physics professor at Clark University in Worcester, Massachusetts. Goddard's rocket was only about 10 feet long and flew only a few miles high. But it proved Tsiolkovsky's theory valid. Goddard subsequently conducted laboratory experiments that showed that rockets could work in a vacuum, which would, of course, be necessary for space travel.[8]

Hermann Oberth, working with a young assistant named Wernher von Braun in Germany during the 1930s, built and launched a number of small experimental liquid-fueled rockets. In 1937 von Braun became technical director of the German rocket research center at Peenemünde. There he developed the V-2 liquid-fueled rocket weapon, the prototype of modern rockets that eventually would be used for space travel. At the end of the war, von Braun was brought to the United States along with other German rocket scientists. He became a technical adviser at the White Sands Proving Grounds in southern New Mexico.

---

*This may have been one of Tsiolkovsky's most prescient observations. While it is now known that humans can tolerate zero gravity for brief periods, it is becoming increasingly apparent to space medicine researchers that long periods without gravity—several months or a year—can have very negative effects on humans. Muscles atrophy, the cardiovascular system weakens, and the immune system may deteriorate. Dr. Harold Sandler, an aerospace physician at NASA's Ames Research Facility in California, said to one of the authors: "What's the point of sending a man to Mars if he can't stand up by the time he gets there?"

This was where the first atomic bomb had been exploded on July 16, 1945, just a few months before von Braun's arrival.

German expertise contributed to the production of a multistage rocket that was launched from White Sands in 1949. It flew 242 miles high, a record at the time. The rocket did not, however, come close to attaining the velocity needed to escape Earth's gravitational bond: about 25,000 miles (40,000 kilometers) per hour. Rocket research, meanwhile, was surging in the Soviet Union. The Soviets had pressed German rocket engineers into service, too, and they had acquired V-2 rockets left over from the war. They also had the blueprint for a monster rocket the Germans had designed to cross the Atlantic Ocean and strike New York City.

After sending a series of bigger and bigger rockets higher and higher, the Soviets prepared to put a satellite into orbit. The command to fire the 100-foot-long single-stage, liquid-fueled SL-1 rocket with Sputnik I aboard probably was given by a man known only as "chief designer." One of the most colorful and mysterious figures in the history of space exploration, he was an engineer whose powerful personality and intellect had turned Tsiolkovsky's theory of mass ratio into the first Soviet space-going rockets. Almost single-handedly, he created his country's space program. His name, Sergei P. Korolev, was kept secret by the Soviets during his lifetime.

Only a little more than a year after Sputnik I, Korolev and his team sent Luna I flying past the Moon. It was the first spacecraft to escape Earth's gravity. A little later a Soviet satellite circled the Moon and sent back the first pictures of the far side. In 1961 a two-stage SL-3 booster rocket carried Yuri Gagarin riding a Vostok spacecraft into orbit. American fears of a dangerously widening gap between U.S. and Soviet science seemed fully justified.

PART II

The Same, Only Different

⋆ 7 ⋆

The Law of Incidental Consequences

The Waters I am entering upon nobody has crossed yet.

—DANTE ALIGHIERI

THE UNITED STATES was all but unprepared for the space age. Its space agency in 1957, the National Advisory Committee on Aeronautics (NACA), had a budget of less than $5 million and a staff of fewer than 500. Its major activity was laboratory research. Less than a decade earlier, the idea of undertaking research in space was ridiculed in Congress and in some U.S. scientific circles. By 1957 space travel still was not taken seriously. The cold war was a fact of life in the United States. Desperation set in as Americans began realizing that the Soviet Union, which was now without question ahead of the United States in space science, might just be a superior military power as well.

The despair was understandable. This was, after all, only a few years after the explosion of the first Soviet H-bomb. The end of the Korean War, which had set U.S. forces against Communist soldiers for the first time, and the ensuing anti-Communist hysteria orchestrated by Senator Joseph Raymond McCarthy and his terrifying permanent investigations Senate subcommittee were still fresh in the nation's short-term collective memory.

Congress founded NASA on October 1, 1958, almost a year to the day after the launch of Sputnik I. The new agency's mission was clear: Make the United States an operational space power as quickly as possible. President Dwight D. Eisenhower named as NASA's first administrator T. Keith Glennan, a former commissioner of the Atomic Energy Commission. Glennan went on leave from his post as president of Case Institute of Technology in Cleveland and moved quickly, merging NACA's Langley, Ames, and Lewis research centers with the Air Force's Jet Propulsion Laboratory, operated by Caltech in Pasadena. The section of the Naval Research Laboratory that had been building the Vanguard research rockets became the Goddard Space Flight Center outside Washington, D.C. Glennan also brought under the NASA wing the Army Ballistic Missile Arsenal (now the George C. Marshall Space Flight Center) in Huntsville, Alabama.[1]

Huntsville's most prominent resident was Wernher von Braun. He had arrived in 1950 from White Sands to become chief of the guided missile development division of the Redstone Arsenal. Von Braun became a U.S. citizen in 1955, and during the 1960s was America's most vocal proponent of rocketry and space exploration (becoming in 1970 NASA's deputy associate administrator). His colorful personality became the model for the title character in Stanley Kubrick's 1964 antiwar film *Dr. Strangelove*.[2] Despite von Braun's enthusiasm and expertise, America's first space efforts were undertaken by another group from the Naval Research Laboratory and were mostly spectacular failures: Vanguard rocket launches that had so far failed to loft into orbit tiny satellites weighing only a few pounds and measuring just a few inches across. The difficulty was that it had been decreed that scientific rockets could not use technology developed by the military, and in particular von Braun had been forbidden to even try to launch a satellite.

At last, in early 1958, before NASA was formed, von Braun and his colleagues at Huntsville managed to launch the United States's first satellite into orbit: Explorer I, a long, thin cylinder that provided the first information about the Van Allen radiation belts. Then the Vanguard finally worked, but American successes still lagged months, sometimes years, behind the Soviets, who by 1959 had managed a hard (crash) landing on the Moon with a spacecraft called Luna 2. NASA finally duplicated the feat in 1964 with Rangers 7, 8 and 9.

The American spacecraft did succeed in sending back more scientific details than their Soviet counterparts, transmitting thousands of pictures of the Moon's surface just before impact.

MISSION TO NOWHERE

In April 1961, Cuban exiles covertly supported by the Central Intelligence Agency unsuccessfully invaded the Bahía de Cochinos (Bay of Pigs) on the south coast of Cuba. John F. Kennedy had been elected partly because of the so-called missile gap, which he claimed existed and which Nixon (correctly, it was learned later) denied in the public debates. Americans were petrified. Relations between the United States and the Soviet Union chilled to freezing. The next month President Kennedy committed the United States to landing men (women were not mentioned) on the Moon (and returning them) before the end of the decade—a bold goal, considering that NASA had yet to launch a human into Earth orbit, although the Moon mission had been pushed by Kennedy's Science Advisory Committee. Kennedy's motivation was political and military, not scientific. Antagonism between the United States and the Soviet Union was approaching a dangerous new level (the Cuban missile crisis, the most terrifying of all cold war confrontations, was only five months away).

Kennedy believed the United States would acquire vast prestige in what had become a global war against the Soviet-engineered spread of communism if his nation could land the first man on the Moon. More important, he wanted to gain the upper hand over the Soviet military in space, since hard-liners on both sides saw the Moon as a logical platform for military operations against targets on Earth. In retrospect, such a fear seems almost frivolous considering the difficulty and expense of getting to the Moon.

The race to the Moon was an invention of the Kennedy White House, one encouraged by science adviser Jerome Wiesner as a means to set a goal that would ultimately establish parity between the Soviet Union and the United States at a time when the Soviets were clearly ahead. Wiesner's reasoning was similar to that used later in establishing the Star Wars program: Choose a goal so difficult that it would require a lengthy, concerted, and expensive effort that only a strong economy such as the United States's could endure. The strategy worked.[3]

The Apollo program that resulted from Kennedy's directive was the largest scientific and technological undertaking in history, one that without another presidential fiat NASA has never been able to duplicate. At its peak in the late 1960s, the Moon-landing program employed directly or through NASA contractors some 600,000 workers and had an annual budget of $20 billion. This was a staggering amount of money, approximately $200 billion in 1995 dollars. The nation's reward for the expenditure came on NASA's first try when Apollo 11 astronauts Neil Armstrong and Edwin E. "Buzz" Aldrin stepped out onto the dusty surface of the Moon on July 20, 1969, as the world listened and watched.

After placing ten more astronauts on the Moon by 1972 and a total of thirty in space, the program ended anticlimactically three years later with the Apollo 17-Soyuz linkup in Earth orbit. Ironically, one of the most successful technological efforts in history, the result of cold war confrontation with a Communist nation, was allowed simply to wither away as funds for it were diverted to help finance a battlefield confrontation with another Communist nation in Southeast Asia.

The Soviets, in the meantime, had tried to duplicate the Apollo mobilization and failed. In 1964 three cosmonauts, the number planned for a trip to the Moon, were launched into orbit aboard a Voskhod spacecraft. During the next Voskhod flight a cosmonaut made the first "walk" in space outside the protection of a spacecraft. In early 1969 two cosmonauts transferred from Soyuz 5 to Soyuz 4, which had linked together in orbit. This was the first time men had moved between separately launched spacecraft and was the kind of procedure the Soviets planned to use for their eventual trip to the Moon.

The Soviet successes, though, were accompanied by spectacular disasters. Dozens of people were killed at the Soviet cosmodrome at Baikonur, southeast of Moscow in central Asia, when a rocket exploded on a launchpad in 1960. In 1967 Vladimir M. Komarov, pilot of the first Soyuz spacecraft, died after a series of mechanical failures caused the capsule to tumble over and over as it returned to Earth. In July 1969, the same month Apollo 11 touched down on the Moon, a huge, supersecret heavy-lift booster test rocket (a later version was called the Energiya), with at least eighteen engines clustered on its first stage that were designed to power the Soviets' lunar mission, blew up. The explosion destroyed an entire launch complex at Baikonur and left scorch marks visible from orbiting satellites. A second test rocket exploded 12 kilometers up, and a third 40 kilometers high. In 1971, after Soyuz 11 depressurized during reentry, killing all three cosmonauts aboard, the Soviets abandoned—permanently, as events were to show—the goal of sending people to the Moon.

With the demise of the Apollo program, the United States, too, abandoned the Moon. Under the Nixon administration, NASA was forced to strike a deal in which it agreed to concentrate all its manned space effort on the new shuttle program as well as to cancel the expendable rockets that had served it so well in the past. NASA officials went along simply in order to continue receiving adequate funding.* During the 1980s NASA struggled to convert what was proving to be a mission to nowhere into a viable Earth orbit program. The USSR, in the meantime, was beginning to prepare for a manned mission to Mars. The Soviets had felt a spiritual pull to the red planet since the days of Tsiolkovosky. Now most of their efforts in space seemed to be aimed at landing humans on the surface of Mars.

Cosmonauts were launched into orbit for stays in the Mir space station of increasingly long duration to test the effects of prolonged zero gravity. The Soviets' plans for an orbital lunar space station and even a colony on the surface of the Moon apparently were designed to gather experience for a three-year round trip to Mars. The logistics of such a trip were daunting. A three-member crew would need five tons of food, nearly twenty tons of water, and ten tons of oxygen. Yet as early as the mid-1980s, the Soviets began planning to send a cluster of three or four Mir spacecraft and service modules with a crew of six to Mars by the end of the century, knowledgeable Western observers believed.[4]

The Soviets did not go far on the journey. They launched a series of unmanned probes toward the planet, two of which missed it altogether

---

* In his book *Challenger: A Major Malfunction* (Doubleday, 1987), Malcolm McConnell maintains that the seeds of the destruction of the *Challenger* space shuttle in 1986 were laid down when NASA was forced to agree to put all its eggs in the one basket of the shuttle program. McConnell forcefully argues that a number of President Nixon's major political supporters were from Utah, where Morton Thiokol, a major shuttle contractor, was based. According to McConnell, there was tremendous political pressure on NASA officials to send as many contracts as possible to Thiokol.

and another two of which crashed on the surface. Two more probes were being planned at the time of the Soviet Union's dramatic dissolution in 1991. They were never launched. Virtually all that survived the breakup of the USSR was a remnant space program with an annual budget of less than $300 million and a vastly diminished technological capacity for space exploration. Symbolizing this demise were two of the world's most powerful rockets, huge Energiyas twenty stories high with engines capable of generating 170 million horsepower. In the mid-1990s they sat rusting in deteriorating launch buildings at Baikonur instead of lifting off with Mir space stations headed for Mars.[5]

By the 1990s NASA, of course, was having its own troubles. After the last Moon trip in 1972, the agency tried Skylab, a short-lived Earth-orbiting station with an astrophysics laboratory and astronaut quarters. The United States did not resume manned space flight until 1981, when the first shuttle was launched. This was designed to ferry people and equipment weighing up to twenty tons into Earth orbit and back. Individual shuttles could be reused and, so it was believed, operate on a monthly basis at relatively low cost. This never happened, and schedule pressures undoubtedly contributed to the explosion of the *Challenger*.

Launched hurriedly on the cold day of January 28, 1986, its rubber O-ring failed, causing a fire that ruptured the main fuel tank. NASA suspended shuttle flights for nearly three years while the spacecraft and its launch system were overhauled.[6] A series of well-publicized failures followed, including that of the Hubble Space Telescope and the loss of a $1 billion satellite. By the mid-1990s NASA was under fire on both the political and media fronts. One article asked, "Is NASA Necessary?" The administration of President Bill Clinton threatened to "reinvent" NASA with huge budget cuts, and some members of Congress began suggesting that the agency be eliminated altogether.[7]

The space race between the Soviet Union and the United States, from the perspective of the 1990s, seemed to be one without a finish line.

SCIENCE MATTERS

Almost unintentionally, the space race was a bonanza for science. In fact, it may prove one day to be an irony of history that the scientific achievements of the Soviet and U.S. space programs are perceived as among the enduring benefits of the cold war. The law of unintended consequences, or at least the law of incidental consequences, apparently worked in science's favor. Rainer Weiss, one of the pioneers in balloon flights to measure the cosmic background radiation, became a scientific adviser for NASA during the years the Apollo program was still sending astronauts to the Moon. He is certain that this is exactly what happened.

"Putting men on the moon had nothing to do with science," recalled Weiss.[8] "Apollo was a macho, cold war thing to do. But they had to have scientists to get to the moon. Once scientists were involved with the moon program, they started pushing NASA to do more science."

Even though never deemed essential for political purposes, scientific research was a major side beneficiary of space technology almost from the day Sputnik I was launched. The first unmanned probes to Earth orbit brought back new details about the cosmic rays bombarding Earth along with facts about the Van Allen radiation belt. Within a few years Luna and Ranger spacecraft were sending back remarkable new data about the composition and geology of the Moon. Luna 9, the first man-made object to soft-land on the Moon, sent back striking pictures that disproved an old (and implausible) theory that the lunar seas were filled with soft, treacherous dust several hundred feet deep. Such probes mapped the entire lunar surface a year before the first Apollo astronauts set foot there.

The six successful manned trips to the Moon (a tragedy was narrowly avoided when the planned lunar landing of Apollo 13 was aborted following an explosion in its service module on the outward journey) brought back some 900 pounds of Moon rock along with literally thousands of new facts about the composition and early history of the earth's only natural satellite: its age (roughly the same as the earth's: between 4.5 and 4.7 billion years), its magnetic fields, its heat flows, and its volcanic and seismic activity.*

Ray Weiss and Dirk Muehlner were approached by NASA to develop a quick experiment that would be interesting, newsworthy, and ready in six months. They devised a thermometer concept to measure

---

*Today one can touch a piece of a rock from the Moon at the National Air and Space Museum in Washington, D.C., and elsewhere.

the temperature of space. The idea required an active impulse cooler stage (to remove the initial heat of the thermometer) from the sensitive thermometer, which would be nested within a set of umbrella radiation shields on the Moon to protect against the Sun but still have a solid angle to deep space. Ray and Dirk suggested that an astronaut set up the experiment while the public watched on television. The experiment was never done, and I suspect that the only "science" most people remember is when an Apollo astronaut standing on the Moon dropped a feather and a heavier object simultaneously. Falling through a perfect vacuum, they arrived on the surface at the same time. This "experiment" confirmed absolutely a 400-year-old theory of Galileo!

When Apollo was over, NASA started looking around for something else to do. Science was an obvious next step, recalled Ray Weiss. During the 1970s unmanned probes plowed the nearly empty seas of the solar system. Mariner spacecraft studied Mars and Venus, Earth's closest neighbors. In 1976 two Vikings landed on the surface of Mars. While their cameras sent back the first pictures of the alien landscape, the spacecrafts' mechanical arms scooped up samples of Martian soil. Automated tests searched for photosynthesis, suggestions of respiratory activity, and metabolism by microorganisms that might be present. All that was found was the *possibility* of organic activity.

Pioneer 10 passed through the asteroid belt, then became the first man-made object to escape the solar system. Voyagers 1 and 2 were launched in 1977 to take advantage of a rare alignment of Jupiter, Saturn, Uranus, and Neptune. The spacecraft passed within 3,000 miles (4,800 kilometers) of each planet's surface, and sent back data about complex magnetic fields and photographs of newly discovered rings that seemed to capture the public's imagination nearly as much as the Apollo flights.

Ray Weiss became directly involved with NASA in 1970. He was born in Berlin in 1932 and came to the United States during World War II with his parents. He graduated from MIT with a degree in physics, then earned his Ph.D. there in 1962. When Penzias and Wilson discovered the cosmic background radiation in 1965, Ray at first was skeptical. Although he knew the technology for finding the cosmic background had been available since late 1940s, he believed that the radio telescope used by Penzias and Wilson was not up to the job. He was soon convinced by the preponderance of the evidence, though, and by the end of the decade he had begun taking his own measurements on the cosmic background.*

By 1970 Ray was regarded in the physics and astronomy communities as a savvy experimentalist adept at quickly resolving technical problems. Along with his young colleague Dirk Muehlner, he was at the time just beginning to make measurements of the blackbody radiation using high-altitude balloons. On their second flight, he was puzzled to find a slight but unexpected excess of this radiation. They discovered that the problem was with their instrument: It was picking up its own radiation, not cosmic radiation. Although they had not been able to prove it completely, this established in Ray's mind that the cosmic background really did fit a blackbody curve.

Their paper set off a controversy of more than twenty years' duration. The issue (which always seems to be *the* issue in science) was who was first—in this case, to establish the blackbody nature of the radiation. Four scientists at Cornell also claimed the distinction: Jim Houck, B. Soifer, J. L. Pipher, and Martin Harwit. Over the years each group has argued that the other's paper was seriously flawed: The MIT results, it was said, had come from a poor instrument, while the Cornell detector allegedly had picked up solar radiation.[10]

The mid-1970s were the heyday of scientific balloon flights. Along with Ray at MIT and his rivals at Cornell, there was the group at Queen Mary College in London led by E. I. Robson and my group headed by Paul Richards at Berkeley. Others like Dave Wilkinson at Princeton dabbled in balloons from time to time. This was because balloons were relatively inexpensive. And at the time they were still the best way to get a handle on the early history of the universe since, flying higher than an airplane, they could get above most of the atmosphere. Of course, they often crashed and destroyed their expensive payloads, and they were difficult to

---

*Ray and I have had a friendly running argument for years over how credit for discovery of the cosmic background was dispensed. He maintains that Ralph Alpher and Robert Herman have received more than enough credit for their role. Ray thinks they did not emphasize the importance of measuring the radiation enough in their papers or subsequent work. I think it is surprising that nobody else did either.

"Alpher and Herman had a lot of ideas, and essentially tossed this one off," Ray has stated. "Their real interest was to look for the mechanism by which elements were formed. And Dicke was too late. He didn't go to the library to do his homework. If he had, he would have found that Alpher and Herman had already made the prediction."[9]

launch even under the best of conditions. They were also a good way to get ready for the more challenging sounding rockets and satellite missions.

The biggest problem, scientifically, was the interference of the ozone layer with the sensitive instruments carried by the balloons. There were only two ways to solve this: Go higher with a balloon, or go all the way into Earth orbit. During the late 1960s and early 1970s, NASA tentatively and somewhat hesitantly had started becoming interested in astronomy and cosmology (which, of course, from a political perspective, was utterly useless). Ray Weiss already had begun encouraging this interest among NASA bureaucrats. He argued that there were new fields of study NASA should look into: planetary sciences, X-ray astronomy, high-energy astrophysics, and particles and fields. Ray had been brought into NASA as an outside science adviser in 1970 through the influence of a woman who worked at NASA headquarters in Washington, D. C., and had became intrigued by his cosmic background experiments.

Her name was Nancy Weber Boggess. She was one of those rarest of bureaucratic creatures: an enthusiast. Unlike most other NASA officials, she was also a scientist. Nancy had grown up in Upper Darby, Pennsylvania, and majored in mathematics at Wheaton College near Boston. After earning an M.A. at Wellesley College, she studied astronomy at Haverford College near Philadelphia, focusing on the probable states of atoms in stars. (When she applied to graduate school in astronomy, she was accepted at Harvard and the University of Michigan. Nancy also had applied to Caltech, but received a long letter from Jesse Greenstein, the admissions director, stating that her "credentials were excellent, but alas, for us, we do not accept women.")[11]

Unlike most young women of the 1990s, Nancy gave little thought to her future as she began her graduate studies at the University of Michigan in Ann Arbor. The first year she met Al Boggess, a laconic young astronomy graduate of the University of Texas, and they married soon afterward. Their first son was born at the end of her second year, and two more children arrived quickly. Nancy, nonetheless, managed to earn her Ph.D., although she recalled that she still was not certain what she wanted to do with her life.[12] In her mind, astronomy was little more than an interesting hobby that she enjoyed when she had the time. On the day her youngest child was ready to start kindergarten, Nancy suddenly realized she had an extra six hours each day.

Maybe now, she thought, she should settle down and pursue her hobby a little more seriously.

Al's career, in the meantime, had taken off. He had gotten a job at Johns Hopkins University in Baltimore, where the family now lived. Al also worked at the nearby Naval Research Laboratory and had become involved with a group of scientists there who were launching rockets at White Sands. These were small, suborbital rockets that would carry a scientific instrument aloft for only about five minutes' or so worth of measurements. Al also sent up sounding rockets over the ocean from a facility at Wallops Island, Virginia.

"It was almost always a comedy of errors in those days," Al recalled. "The rockets usually weren't hermetically sealed very well, and the payloads would sink when they came down."[13] Once he and his colleagues tried to photograph a solar eclipse with a camera mounted on a sounding rocket. It, too, sank. Fortunately, divers were able to recover it.

Because of his rocket expertise, Al was offered a job at NASA in 1959. In 1961 he organized a NASA expedition to the Woomera Rocket Range in the Australian outback. This was a military facility operated by the British which Al and other NASA scientists used to launch rockets to photograph the southern sky. By the mid-1960s Al and other scientists within the space agency were agitating for NASA to do still more science—particularly since Earth orbit was such a perfect place to undertake stellar or solar observations. In 1965 NASA launched its first orbiting astronomical observatory into orbit. The flight was a disaster because the payload was improperly engineered, Al recalled. The first astronomical instruments were carried into orbit in 1967 aboard Scout rockets. These were relatively small, solid-fuel rockets that could be mass-produced. They were so easy to launch that they could be hauled virtually anywhere and shot off like firecrackers.

Astronomers quickly learned that the Scout was of little use for astronomy because astronomers' main instruments were telescopes and other large detectors that had to be pointed accurately in one direction or another. The Scout was just too small to carry such a large payload into space. Looking around for something bigger, Al and other NASA scientists started pushing the agency to fire scientific payloads into space aboard medium-sized Delta rockets. These originally were conceived by Bill Schindler to carry a payload of several hundred pounds into orbit using the Air Force Thor as a first stage. Moreover, larger engines and new stages could be added to a Delta, meaning that it had the advantage of being able to grow just like an Erector set as technology evolved.

About this time astronomical research at NASA received a real boost, although nobody knew it at the time, when Nancy Boggess resumed her career in earnest. In January 1968, she gave a talk at the American Astronomical Society meeting in New York on NGC 6822, a dwarf, irregular galaxy. Dr. Nancy Roman, a science administrator at NASA headquarters, happened to be in the audience. Impressed by what she heard, she offered Nancy a job in her office in Washington. The job would be to oversee progress on fifty or so NASA research grants. It would be only half-time at first, which would give Nancy plenty of time to spend with her children. She jumped at the chance.

By the early 1970s scientists outside the agency were also clamoring for access to Earth orbit. To accommodate the growing demand, NASA headquarters in Washington came up in 1967 with the idea of issuing an "announcement of flight opportunity" whenever a Scout or Delta rocket was to be launched for the purpose of scientific research. The idea, of course, was to spread around the unique vantage point of an orbiting instrument whether it was for studying the earth, the solar system, or the cosmos.

Nancy Boggess and Ray Weiss eventually were to play major roles at NASA headquarters in determining how these "AO's" were implemented and how the flight opportunities were awarded. This meant, of course, that they would have a direct effect on the proposal I was planning to submit for a blackbody radiation experiment in space.

✳ 8 ✳

The New Aether Drift

I have little patience with scientists who take a board of wood, look for its thinnest part, and drill a great number of holes where drilling is easy.

—ALBERT EINSTEIN

RAY WEISS WAS THE FIRST person I called in 1974 as I began gathering a team of scientists to submit a proposal to NASA for a cosmic background experiment in space. I was still working as a postdoc with Pat Thaddeus at the Goddard Institute of Space Studies in New York City. Ray's name was on the list Pat had given me to call. "If you can get Ray to sign on, your chances with NASA will be better," Pat told me. I already knew about Ray because of his balloon work on the cosmic background spectrum. He was a top experimental physicist, and highly respected within the physics community. His accidental discovery—that the small window for the far infrared instrument was not always needed—was famous within the balloon community.

He also had learned that infrared instruments could suffer from the so-called Narcissus effect. This happened when the detector radiated heat. This heat would then be reflected back from another part of the instrument to be measured as if it had come from a cosmic source. Moreover, Ray had helped develop atomic clocks and had become directly involved with NASA, having served since 1970 on its Physical Science Committee (of which Robert H. Dicke was chairman). He was on two of the agency's working groups, one for airborne astronomy and the other for shuttle astronomy. He was chairman of the NASA Panel on Experimental Relativity and Gravitation.

Pat knew more than I the importance of such contacts. But I agreed that Ray would be perfect for a cosmic background team. But would he agree with our assessment? I called Ray, introduced myself, and asked if he would be interested in joining a space venture aimed at taking more precise measurements of the $3°$-Kelvin radiation. Ray told me rather bluntly that he would not: "It's too problematic. And it's too expensive. I've got too many irons in the fire right now. You'd better go find somebody else."

I decided to go see him. Boston was not far from New York, and I believed that if I could show him my rough idea for a sensitive new detector, he might relent. Perhaps he was more intrigued than he admitted. I called again, and he agreed to meet with me. I rode the train up to Boston and took a cab over to MIT across the Charles River in Cambridge. Ray's office was in a warren of rooms on the second floor of Building 20, a decrepit structure facing Vassar Street on the north side of the campus. Dating from the time the United States entered World War II and built of wood with wings added haphazardly over the years, it looked like a factory from Dickensian London rather than one of physics' most storied sites.

Building 20 was one of the places where radar had been perfected during the war. The wings of the building were added as the role of the radar lab changed from simply research to actual construction of equipment. People working there felt the hot breath of the war directly, since there was no time to go through the full cycle of design engineering, construction, and test of the special-purpose radar systems that were needed immediately. Equipment fabricated in Building 20 was operating on warships and in airplanes only a few days after the last wire was soldered into place.

Building 20 was also where Robert Dicke had built the first Dicke radiometer, then climbed up on the building's roof to take what could have been the earliest reading of the cosmic background radiation in 1946 (but which had not been recognized as such at the time). Most of the physicists who work there cherish the old building despite its long, dark hallways (frequently leading to dead ends) and its lack of amenities such as central air conditioning. MIT officials have threatened on occasion to raze the building, but the physicists always manage to save it.

Ray Weiss loves the old building. Like the structure, he has little use for formality. It is not unusual for him to work there in his undershirt, particularly in the summer. We met in his office, surrounded by teetering bookcases and rickety tables laden with stacks of books and papers. I was impressed immediately. Ray had been named a full professor of physics the year before, but he was a down-to-earth man who often had grease under his fingernails from working on his old Volkswagen Beetle. He made experimental electronic equipment himself, milling, drilling, and soldering the parts in his laboratory. He loved to teach his students how to work with their hands. He was full of zip, vigorously smoking a pipe that he carried in the pocket of his tweed sports jacket along with a tin of tobacco. He played the piano and loved music and, I learned later, was a romantic at heart. He definitely was my kind of person. The real question, though, was whether I was his. We talked over my ideas for a proposal for two or three days, and Ray remained skeptical.

"The cost is going to kill you in the end," he said. "Do you have any idea how much money you're getting into?"

It was a good question to which he already had the answer: of course not. None of us did. He also was worried that the Narcissus effect (in which the detectors see their own reflections) could ruin the data taken by a satellite-borne instrument unless the instruments could be kept cold enough—just a few degrees above absolute zero. He considered this to be next to impossible on a satellite. Most of all, he was worried that the dipole anisotropy had not yet been detected conclusively.

This was an effect that Ray and other cosmic background researchers believed would be caused by the motion of the earth relative to the radiation. The idea grew out of the concept that the cosmic background radiation was a kind of universal frame of reference against which might be detected absolute motion of the earth, the Milky Way, or maybe even the entire universe. This idea was an old one in physics: Newton had suggested that space and time, which he believed were both absolute in nature, would provide such a universal matrix.

Reviving an ancient Greek idea that heavenly bodies were propelled through space by an idealized, crystalline "aether," nineteenth-century physicists combined the old aether concept with Newton's belief that space and time were the only universal frame of reference. But instead of a crystalline substance that carried the planets and stars along in its wake, the aether was now a virtually undetectable substance through which light waves traveled. Victorian physicists searched long and hard to find evidence of the aether. In what undoubtedly was one of the most famous nondiscoveries in the history of science, Albert A. Michelson, a physics instructor at the U.S. Naval Academy in Annapolis, joined forces with Edward W. Morley, a renowned American chemist, to track it down.

Leading theorists on the nature of light such as Augustin Jean Fresnel and Sir George Gabriel Stokes had reasoned that as the earth traveled through the aether at 18.6 miles per second on its trip around the Sun, the planet's motion would stir up an "aether wind." And if light waves propagated through the wind, then the aether must offer at least *some* resistance to them. Hence light traveling against the aether would be expected to move more slowly than light traveling with the aether—just like a canoe in a river goes faster downstream than upstream against the current.

In order to detect the aether, it was necessary to know the velocity of light. Galileo had first tried to measure its speed in an experiment in which he stationed two assistants on distant hills and had them flash lanterns at each other so he could time the interval. Without access to a clock, Galileo tried to use his pulse to time an interval that has since been estimated at about one hundred-thousandth of a second! Michelson had better luck in 1877. The son of a Polish immigrant merchant, Michelson had graduated first in his class from the Naval Academy four years earlier. He was very bright, and a whiz at optics.

His experiment was simple and inexpensive, consisting only of a lamp, a condensing lens, and two mirrors 500 feet apart. One mirror rotated at 130 revolutions per second, while the other was fixed. Light focused on the rotating mirror was reflected through the lens onto the fixed mirror, then back again in flashes. The amount of displacement could be measured accurately as an angle, which allowed him to calculate the speed of his light flashes at 186,508 miles per second—a remarkably accurate number.*

---

*In 1907, Michelson became the first American to win the Nobel Prize for physics. He was cited for the "exactness of measurements" relating to light. Several years later Einstein paid tribute to him: "Through your marvelous experimental work [you] paved the way for the development of the theory of relativity. Without your work this theory would be scarcely more than an interesting speculation."

Working on the aether problem with Morley a decade later, Michelson built an extremely sensitive optical interferometer. This timed the mirrored return of two simultaneous beams of light, one fired at right angles to the aether wind and the other directly into it. Beginning in 1887 Michelson and Morley tested for differences in c, the velocity of light, into the aether breeze and across it at all angles. The results were always the same: nothing, no evidence whatsoever of an aether wind. They tested and retested their apparatus, the strain of constant failure eventually leading to a nervous breakdown in the obsessive Michelson.[1]

At last they—and the thousands of anxious scientists around the world who were following the experiment—realized the terrible truth: There was no aether wind. And, by extrapolation, there was no aether, either. Michelson and Morley had revealed a startling anomaly in Newtonian physics. Yet the meaning of their results was not clear. The ensuing crisis in the Newtonian paradigm of absolute time and absolute space lasted for nearly two decades and was not resolved until Einstein proposed his special theory of relativity in 1905.

The aether concept was intriguing because it seemed to share some qualities of the cosmic background radiation. Dennis Sciama, a cosmologist at Cambridge University (and Stephen Hawking's former mentor there in the 1960s) had suggested this in 1967. Sciama believed that, like the aether, the cosmic background radiation could be used as a kind of universal frame of reference.[3] After all, the background radiation pervades the entire universe, Sciama argued, and is supposedly homogeneous in every direction. If careful measurements could be taken of this background radiation, the motion of the earth or the Milky Way in relationship to distant parts of the universe might be detectable.

How could this be possible, though, if the cosmic background were the same everywhere? What could be measured? The radiation was like a white fog suffusing all of space. Measuring the background radiation was like looking into such a fog, the same everywhere with no identifiable markers anywhere. Sciama believed, though, that there would be a slight variation in the temperature of the cosmic background radiation depending on the direction of our motion through it. This was due to the Doppler effect.

The Austrian physicist Christian Doppler discovered this shift in 1842 in the context of sound waves. As a whistling train approaches, the frequency—or pitch—of its whistle rises. As the train departs, the whistle's frequency drops. For light, the Doppler effect would affect its wavelength, shifting it toward the blue end of the spectrum for an approaching object and toward the red end for a receding object. Edwin Hubble had identified the redshift of distant galaxies and concluded that they were receding from us. Sciama believed the same principle would apply to our movement through the cosmic background.

If the earth were not traveling through the cosmic background radiation in relationship to other parts of the universe, the cosmic background would appear to us to have *exactly* the same temperature in every direction we measured it. It would be *isotropic*. Sciama knew, of course, that the earth moved with respect to the rest of the universe: in its orbit around the Sun, within the Galaxy, and in relationship to distant galaxies. He reasoned that the Doppler shift would make the radiation's temperature appear slightly higher—on the order of about one part in a thousand—in the direction of our motion. Conversely, the temperature would be slightly cooler in the direction of recession. In other words, the temperature would be said to be *anisotropic,* or variable according to the direction in which it was measured.

In 1971 Jim Peebles, following up his role in the discovery of the cosmic background radiation, published a small book called *Physical Cosmology* that became a sort of bible to my generation of observational astronomers and cosmologists.[4] In the book Peebles explained the significance to cosmology of the cosmic background radiation, citing the early theoretical work of Ralph Alpher and Robert Herman. In a section called "The Aether Drift Experiment," he pointed out the similarities shared by the Victorian aether and the cosmic background radiation.

Each was, in a sense, a modern manifestation of Newton's concept of space as an absolute frame of reference. Although this was true for the aether and the cosmic background radiation, they were nonetheless radically different. Where the aether required as confirmation of its existence a variation in c, the speed of light, the cosmic background radiation had no such requirement. Hence the radiation as a frame of reference did not violate Einstein's theory of special relativity.

"It always is possible to define motion relative to something, in this case the homogeneous sea of radiation," Peebles stated.[5] He calculated the degree of warming and cooling in respect to the motion of the earth as it plowed through the radiation sea. Knowing that our solar system orbited the galaxy at a velocity of approximately 250 kilometers per second, Peebles showed the amount of anisotropy between the warm pole in the direction of forward travel and the cold pole in the backward direction: about 0.08 percent. This was almost one part per thousand, approximately the amount Dennis Sciama had predicted.

The cosmic background radiation would provide a perfect observational laboratory, Peebles believed, for studying the relative motion of the earth, our home galaxy, distant galaxies, even the entire universe. This latter idea, that the universe itself is rotating, was proposed in 1947 by Kurt Gödel, the remarkable Czechoslovakian-born mathematician who was Einstein's friend and colleague at the Institute for Advanced Study in Princeton.[6] Gödel based his finding on a new kind of solution he had worked out to Einstein's field equations of general relativity. Gödel's new solution also included the tantalizing hypothesis that, in a rotating universe, travel through time might be possible.

If the entire universe were rotating, as Gödel proposed, one effect of such motion should be that measurable temperature variations would appear in the cosmic background radiation. The equator and the poles of the rotation ought to be recognizable as hotter or colder. These variations, of course, had never been seen (and still have not). A good first step, though, for observational cosmologists would be to find the dipole anisotropy. This is what physicists call the effect produced by the Doppler shift as the earth moves through the cosmic background radiation: *dipole* because the background radiation would have two poles, one hot and one cold, and *anisotropic* because the temperature of these poles would not be the same.

LACK OF EVIDENCE

Ray Weiss was deeply concerned and frustrated that by the mid-1970s conclusive evidence of dipole anisotropy had not been found.

"I'm completely flummoxed by it," he admitted when we first met in 1974.

"We're seeing junk everywhere," he said, referring to hot and cold spots in the background radiation caused by the galaxy, the Earth's atmosphere, or the detecting instrument itself, "but no dipole."

Two groups had detected tentative evidence of the dipole.[7] But Ray was not yet convinced. He believed it was an essential first chore in the hunt for anisotropy, if it existed, in the cosmic background radiation. Along with the lack of conclusive evidence for the dipole, Ray also was worried that the single instrument that I was proposing to launch into orbit aboard a satellite to refine the blackbody spectrum line would not be up to the job. There simply would be too much interference from numerous other sources of radiation in the universe, Ray believed. "Unless you have multiple coverage, the galaxy is going to fuck you to the wall," he said.

His idea for developing several different kinds of instrumentation in order to eliminate, or at least cancel out, possible interference from radiation sources such as the galaxy and the Sun seemed like a good one, I agreed. But as we came to the end of our meetings, Ray was still not convinced of the feasibility of my proposal.

"Go see if you can get Frank Low and Dave Wilkinson on board," he told me. "Then come back, and we'll see about it."

An astronomer at the University of Arizona, Low was well known for his development of new detectors capable of seeing far infrared radiation, or radiation with wavelengths up to 1 millimeter. This was useful for looking at very cool sources of radiation only a few degrees above 0° Kelvin, such as would be found when measuring the total energy output of our galaxy or others. Some of the visible light emitted by individual stars in a galaxy would very likely be absorbed by interstellar dust and then re-emitted at the far infrared wavelengths.

Low at first had tried out his new detectors in conjunction with ground-based telescopes, but the air near the ground was troublesome. He then mounted his detectors along with a smaller telescope on NASA's modified Lear jet, based at Ames Research Center in California, in order to get above most of the atmosphere. This was a heroic and pioneering effort. But he was just scratching the surface. His telescope was simply too small, only 12 inches in diameter. Only a few cosmic sources were bright enough for it to measure. Frank did not join our team. He was in the process of helping submit another proposal to NASA for an infrared astronomical satellite.

Dave Wilkinson, of course, I knew personally dating back to my oral exam at Swarthmore six years before. I also by now knew him by reputation for his involvement in the discovery of the cosmic background radiation. Dave is by nature skeptical and reticent. He was busy with his own work, he told me when I called. For one thing he was in the midst of planning with a Princeton colleague, Brian Corey, a new search for the dipole using a detector mounted on a high-altitude balloon. Moreover, he was not sure he was interested in working on a big space project that promised only endless committee meetings and results that might not appear for years. Despite his reservations, I felt strongly that Dave was the right person for us. I told him about my work with Pat Thaddeus and my talks with Ray Weiss.

"Let me think about it," he said. "Let's stay in touch."

Pat also had suggested I call Mike Hauser at the Goddard Space Flight Center run by NASA in Greenbelt, Maryland. Although new at Goddard, Mike was a key person to call. Mike had been brought into the NASA fold by Frank McDonald, father of high-energy astrophysics at Goddard, to build up a small infrared astronomy group. Mike had graduated from Cornell, received his Ph.D. from Caltech, and taught physics at Princeton, where he had been involved in high-energy particle physics before shifting over to astrophysics. Later at Caltech he helped develop bolometers for infrared astronomy. So he already was aware of the value of space for doing further astronomical observations.

At Goddard Mike had organized a balloon research program to develop infrared observational techniques. Mike's group included only one other scientist so far, Bob Silverberg. Bob also was a converted high-energy physicist who had been heavily involved in studying cosmic rays. He had earned his Ph.D. working at Goddard and the University of Maryland in nearby College Park. Mike said Bob would like to work with us on the proposal. He sounded like a fine addition to me, and I agreed at once. I didn't know it yet, but Mike was a real dynamo who not only would work on our proposal but also organize another entire group at Goddard to submit a competing proposal for an infrared astronomy satellite. This proposal eventually landed Mike on the infrared team with Frank Low, which was to prove very valuable to our group.

Pat Thaddeus, in the meantime, decided he was interested enough in the project to put his name on the proposal. I was delighted. Pat was well known throughout the astrophysics community for his work in radio astronomy, having discovered a vast number of interstellar molecules.

Inspired by his own work on the excitation of interstellar cyanogen molecules (which itself had been inspired by the discoveries of Andrew McKellar back in the 1930s and 1940s), Pat had written a review article about the way the molecules could be used to measure the temperature of the cosmic background radiation. Pat would bring some expertise to the project. I also thought his name on the proposal would give us a better chance of approval.

At the same time, I realized it was unlikely that he would remain as a permanent member of the team. Pat was all too aware of the long time frame for attaining observational results in space. He preferred the shorter-term results of radio astronomy. One person I did not call was Paul Richards, my old thesis adviser at Berkeley. Some of us discussed the wisdom of asking Paul to join us, since we all realized he would be an excellent addition to the team. He had done fine work on the spectrum and he was highly regarded.

"If you call in Paul, everybody will think of him, and not you, as the leader of the project," people told me, advising against it. We had also heard through the grapevine that Paul was not especially keen about the idea of a space project. In any event, I never discussed it with him directly. We were concerned that if Paul heard about our plans, he might mention them to other physicists or astronomers at Berkeley. They could eventually compete with us for the spot on the NASA satellite, or Paul himself might lead such a group. As it turned out we did eventually face competition from Berkeley, but from another group entirely.

As the team took shape, we began contacting other people who might be able to help us. We learned that Ball Brothers Research Corporation in Boulder, Colorado, had already begun designing a dewar,* a liquid helium cryostat for the Infrared Astronomical Satellite, which the Netherlands planned to launch five years later, in 1979 (it finally went up in January 1983). Our instruments also would need one of these. It was a kind of giant Thermos bottle for maintaining a constant low temperature that would prevent a detector from reading its own reflected heat. We called in Alan Sherman, a Goddard expert on space cryogenics, to help us with its preliminary design. Other engineers at

---

*Dewar vacuum vessels are named after Sir James Dewar (1842–1923), a Scottish physicist and chemist who invented them shortly after the turn of the twentieth century.

Goddard gave us a quick series of tutorials on what was possible in terms of designing a space mission. LTV Corporation of Dallas, which built the Scout rocket we thought we would be using, sent an engineer named John Pacey up to New York to fill us in on the details of how the rocket was launched.

In July I took the train down to Washington to attend a pre-proposal briefing at Goddard, and in September filed the requisite "notice of intent" with Frank Gaetano at NASA headquarters in Washington. During this period I received two pieces of good news: Ray Weiss called from Cambridge to say he wanted to join the team, and would bring along his fellow balloon experimenter, Dirk Muehlner; and Dave Wilkinson signed on at the last minute, although still somewhat reluctantly because of his concerns about getting involved with a long space project with NASA.

The proposal was due at the end of October. There was little time to dawdle.

A PLATFORM IN SPACE

We met together for the first time on September 27, 1974. Pat chaired the meeting, which took place in the second-floor conference room of the Goddard Institute for Space Studies in New York City. The institute occupied the floors above Tom's Restaurant in an old building on 112th Street and Broadway. Because of the anti–Vietnam War demonstrations and riots at nearby Columbia University just a few years earlier, all signs indicating the institute's presence there had been removed. The building had a guard at a desk where we had to show our badges every day.

Sitting around the conference table with Pat and me were Ray Weiss and his young colleague Dirk Muehlner from MIT, Dave Wilkinson from Princeton, and Mike Hauser from Goddard in Maryland, along with Joe Binsack from the MIT Center for Space Research. Although he intended to work with us, Bob Silverberg did not make the trip up from Washington. Everything considered, it was a rather strange group, all white male physicists, fairly young (I was the youngest, at twenty-eight) to early middle age.

Despite the fact that we were planning a space venture aimed at taking new measurements of the universe, none of us had been trained as a cosmologist. And only one of us—Joe Binsack, who would soon drop out—had been even slightly involved in launching scientific hardware into space. We knew nothing of the years of hard work, the many difficulties and heartbreaks ahead of us. Our naïveté undoubtedly was both a blessing and a strength.

Each person there related what he knew about existing science projects in space and the technology used. In order to keep our instruments cold enough to produce meaningful results, we knew that we were going to have to send liquid helium into orbit for the first time (for Americans). This was a big worry to everybody. What would happen as the exceedingly cold liquid—just a few degrees above absolute zero—sloshed around inside a dewar during launch and once in orbit? Ray Weiss said he had heard about an Air Force satellite that had carried a helium dewar into space.

"But it wasn't carrying a liquid," Ray said. "It was only supercritical."

Helium gas had been chilled to about $5°$ Kelvin at high pressure and, although nearly as dense as a liquid, was still a gas. Ray also mentioned the Air Force's Celestial Mapping Project, which had used a mechanical refrigerator capable of chilling a gas or liquid down to about $15°$ Kelvin. There was a rumor that the Soviets had launched a satellite carrying a helium-filled dewar, but they were not likely to share the details with us. A more hopeful source of information might be a cryogenics group at Stanford University led by Bill Fairbank and Francis Everitt. They planned to launch an extremely sensitive gyroscope with a liquid helium–filled dewar in order to test Einstein's theory of general relativity (and are still working on it at the time of this writing). Ray knew they already had invented a crucial item: a porous plug that separated superfluid liquid helium from the vacuum of space.

Nobody had any idea of the difficulties we would face in trying to send liquid helium into space. But everyone there recognized the incredible advantage our instruments would have once they reached orbit. Despite the worries about the dewar, the empty environment of space would provide a remarkably clean, controllable place where it would be relatively easy to keep instruments at an even temperature. There would be no atmospheric noise or weather problems that could disrupt a balloon flight to worry us, and no freezing or condensing water.

Satellite-borne instruments would have a long working life compared with those aboard balloons or high-altitude planes. This would allow us plenty of time to calibrate the instrument, take measurements, and then redo the measurements if necessary. The satellite would spin slowly in its orbit around the earth, giving us a clear picture of the sky in every direction. It was an exciting prospect, everybody seemed to agree. Dave Wilkinson was the single exception. He already had begun taking what would become a customarily skeptical, occasionally negative—and, as I was to learn, almost always necessary—contradictory position.

"I want to do good science. I don't care if it's from a balloon, if it's on the ground, or if it's on a satellite out in space," Dave said. "I'm not very interested in NASA's bureaucratic intervention from the day we start to the day we finish."

In spite of his doubts, Dave agreed to go along with our plans to write a proposal. But he indicated that if the NASA committee meetings and layers of bureaucracy became too much for him, he simply would drop out. He never did, but he never enjoyed the committee meetings either. He later had plenty of his own when he became chairman of the Physics Department at Princeton.

With the proposal deadline only about a month away, we faced a crucial question: What kind of detectors would we tell NASA we were going to send into space? Ray Weiss insisted on several complementary instruments. "You've got to do it that way, otherwise galactic noise and everything else out there is going to screw up your data," he said, reiterating what he had told me in Cambridge several months earlier. Ray and Dave pushed for a cosmic background mapping experiment.

This would be accomplished with a set of microwave radiometers with flared-horn antennas at four wavelengths: 3, 5, 9, and 16 millimeters. Ray thought we could obtain a sensitivity of 1 part in 10,000 to 1 part in 30,000. As we learned much later, this would have been nowhere nearly sensitive enough for what we had to accomplish in order to see the cosmic background lumps and bumps. Like all the instruments in the proposal, this one would be modified substantially as we worked out the details.

A second instrument Ray wanted was a modification of the balloon-borne radiometers he and Dirk had used to measure the anisotropy of the cosmic background radiation in the far infrared

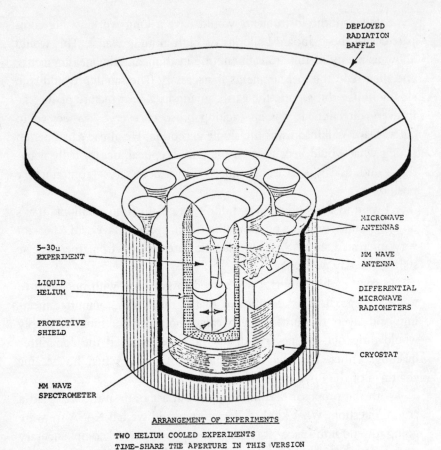

Fig. 2. JOHN MATHER'S ORIGINAL DRAWING FOR A
COSMOLOGICAL BACKGROUND RADIATION SATELLITE.
This first drawing, less one instrument, was remarkably like the COBE satellite
that eventually was launched into orbit.

wavelengths. This instrument would, Ray believed, give us a different
look at the sky than the view with the microwave radiometers at a
much greater sensitivity: something on the order of 1 part in 100,000
with an angular resolution of 5°. Angular resolution is a measure of
how sharp our images would be, and 5° is still pretty fuzzy. The
smallest objects we could recognize would be the size of a fist at arm's

length. The instrument would have a sort of motorized chopping wheel on the outside of the crysotat. This would allow the radiometer inside to look alternately at signals coming in from two different antennas.

Our third instrument would be a polarizing spectrometer with a horn antenna aimed at working over a broad range of wavelengths. A flared section of the horn that looked like the bell of a trombone would suppress diffracted light coming from the wrong directions. This was the idea I had developed as an improvement of the double-cone antenna we had built for the balloon flight with Dave Woody and Paul Richards at Berkeley. By then Woody and Richards and Norm Nishioka back at Berkeley had fixed up our payload and flown it successfully. So we had real data to look at, and I began feeling optimistic.

The satellite instrument would be designed to make a far more sensitive measurement of the blackbody spectrum than ever would be possible at a suborbital altitude. The detector would use a rotating filter wheel to select various wavelength bands for study, along with a rotating chopper wheel to modulate the radiation. Although the spectrometer design was new—technically, a differential photometric comparator with a full-beam external blackbody calibrator—it was a classic concept. Everybody believed that if it worked at all, it would work very well.

Mike Hauser pushed for one more instrument. This would be a rotating filter wheel in front of a few detectors receiving light from a 20-centimeter aperture telescope. Since the late 1960s Jim Peebles had pushed for such an instrument to measure the cosmic infrared background radiation from the earliest galaxies. We decided on wavelengths from 8 to 30 microns. As a side effect, the instrument would also be sensitive to foreground noise: infrared sources such as stars, interplanetary dust, interstellar dust. We knew it would be tough to tell the signals apart—and we were right.

Even if any of us had been old hands at sending experimental hardware into space, the suite of four instruments we had come up with was incredibly ambitious. The fact that we were proposing to load all of these detectors aboard a cryogenically cooled satellite was probably a completely crazy idea. In retrospect, the craziest thing about it was that we thought we knew what we were doing and what we were getting into. I tried not to worry about it: If our proposal was selected,

NASA probably had the expertise and talent to pull it off. We did not have to have all the answers just yet.

I drew a rough sketch with the instruments jam-packed in a satellite. Complicating our proposal further was that our satellite would spin, changing direction continuously in order to obtain views of different parts of the sky. As far as anybody knew, nothing like it had ever flown in space before. It would not be easy to get the torque to slew continuously around the spin of what essentially was a huge gyroscope. But I knew that all we needed was a counter-rotating wheel inside the satellite with equal but opposite spin angular momentum. With the gyroscopic forces of the two spinning objects canceling out, it would be easy to point the spacecraft in any direction we wanted.

Our proposal contained several innovative ideas. But leaving the meeting, I wondered whether it would stand a chance against the proposals submitted by more senior groups. On the other hand, if it were approved and were to work, it would be wonderful, I thought. It would give us an entirely new set of pictures of the conditions in the infant universe.

As the proposal process went on over the coming months, we learned that two other groups, both in California, had come up with very similar ideas about how to measure the cosmic background radiation from space.

✳ 9 ✳

Parallel Universes

It is an hypothesis that the sun will rise tomorrow: and this means that we do not know whether it will rise.

—LUDWIG WITTGENSTEIN

PEOPLE ON OUR PLANET now, as in every age, can look up at night and behold a magnificent black canopy filled with bright planets and stars blinking like lighthouses on a distant seashore, as well as the familiar constellations of the Big Dipper and Orion. The Moon can still inspire fear as, huge and orange, it rises above the horizon. If the air is clear and cold and city lights far away, you can still pick out the faint fuzz of the Andromeda galaxy two million light-years away. Looking into the sky and trying to fathom the secrets it holds is an experience that has been relived countless times. But the picture of the universe framed by the night sky, we know now, is too simple.

Today's astronomer observes a universe far different from that viewed by any predecessor. Four centuries ago, rudimentary telescopes were first trained on the night sky. By the beginning of the twentieth century, astronomers had at their disposal light-gathering instruments sufficiently powerful to unveil the true vastness of a cosmos that to their astonishment was constantly changing. Since then, many new kinds of detectors have been added to the astronomical arsenal. With their sensitivities improving each year, these instruments have unveiled entirely new regions of the sky filled with X rays, infrared radiation, gamma rays, ultraviolet light, or radio waves—the universe no longer a single universe but a set of parallel universes.

Pick one of the new instruments. Imagine you can look through it like an old-fashioned telescope or pair of binoculars.* Look at the sky. It is utterly black. You see nothing you recognize, for now you are looking into the gamma ray universe. It appears to be empty. Suddenly there is a brilliant pinprick flash, gone in an instant. Then nothing for hours on end. At last, another flash comes from someplace else in the sky. After a while you see another from somewhere else and, eventually, another. These are gamma ray bursters. Gamma rays are the most energetic form of light anywhere and the kind with the shortest wavelength, ranging from about 10^{-11} to 10^{-13} meters. On Earth they can be created by nuclear reactions, often accompanied by alpha and beta rays, or by a radioactive substance as its nuclei change to a lower energy state; gamma rays also can be made by collisions of high-energy electrons with targets in accelerators. Astronomers are utterly perplexed by gamma ray bursters. They remain one of the great unsolved mysteries of the cosmos.

Try another kind of astronomical detector: one tuned in to detect X rays, the next step up the electromagnetic spectrum. What do you see in this new universe? Gone are the familiar patterns of the night sky. You are now looking out into a universe that is highly energized. Here and there are intensely bright spots where matter in the form of gas is being swallowed up by black holes or where stars are combining into greater and hotter stars. X-ray astronomers analyze these intensely hot objects by studying the X rays they emit, usually in the wavelength range from 0.01 to 10 nanometers (1 nanometer $= 10^{-9}$ meter). Invisible to ground-based telescopes, astronomical X rays were discovered in 1949 when detectors launched on V2 rockets found them pouring out of the Sun. In 1962 R. Giacconi led a group that launched a small rocket equipped with detectors similar to Geiger counters.

During a five-minute flight, the instruments picked up X rays from a nearby binary star system now called Scorpius X-1, and discovered that diffuse X rays bathe the earth. An early joint NASA-Italian astronomical satellite called Uhuru ("freedom" in Swahili), launched by a

---

* Astronomers do not look through any instruments today, even telescopes. The instruments collect data that observers read and then analyze, almost always by computer.

Scout rocket in 1970, succeeded in mapping more than four hundred sources of cosmic X rays. The most startling was a series of binary stars in which ordinary stars orbit neutron stars emitting X rays. One called Cygnus X-1, an object with ten times the mass of the Sun, was far too massive to be a neutron star. Astronomers think it could be a black hole.*

Pick up another instrument. The sky is more unrecognizable than ever. The Moon is gone. So are the planets and the familiar stars. In their place are clouds of white light. Unfamiliar stars appear here and there. Your detector is registering the ultraviolet universe, one that shines invisibly to the naked eye with the kind of electromagnetic radiation that lies between visible violet light and X ray; its wavelengths lie between 10 and 400 nanometers. Some of the Sun's radiation is ultraviolet, most of which is absorbed by the ozone layer in the upper atmosphere (at least for the moment). Hotter stars produce vast amounts of ultraviolet light, as do white dwarfs, the naked cores of old stars that have shed their protective shells.

Now try a familiar detector: your eye. Enhance its distance vision with a telescope, and you are in the universe of optical astronomy. Observers in this oldest branch of astronomy detect and study light sources throughout the universe: celestial objects such as planets and their moons, which reflect light from the Sun, and stars and galaxies that emit light on their own, their visible light waves covering a very narrow wavelength range, from about 400 to 700 nanometers (4,000 to 7,000 Angströms).

Peer through another kind of detector. Again the familiar night sky disappears. The bright, comforting glow of the Moon is oddly different, even disarming. In the infrared universe all you see is the heat the Moon itself emits, heat that it once absorbed from the Sun. The infrared universe is a cold one, filled not with hot objects but relatively

---

*In 1978 NASA launched the High Energy Astrophysical Observatory II (HEAO II), known also as the Einstein Observatory, which detected several thousand new X-ray sources in the Milky Way and beyond. It also discovered that cataclysmic variable stars in our galaxy emit X rays when they are in outburst. NASA was building in the mid-1990s an advanced orbiting observatory called Advanced X-ray Astrophysics Facility (AXAF). It will be fitted with a 1.2-meter mirror and will be ten times more sensitive than the Einstein Observatory. Others are in orbit or in preparation.

cool ones emitting electromagnetic radiation with wavelengths in the range of 0.0001 to 0.1 centimeters. A few stars blink on and off over the course of a year or so in the infrared universe, very strange ones. Known as cold stellar bodies, they either are red giants dying or new stars being born. First detected by Newton as he examined sunlight with a prism and a thermometer, infrared radiation is thermal. Today's infrared astronomers know that thousands of objects populate the infrared universe: stars with low surface temperatures, sources in Cygnus, the Orion Nebula, and at the center of the Milky Way, our own Sun. Highly sophisticated infrared detectors provide astronomers with information about planetary atmospheres, stellar evolution, and galactic structure. Because the earth's atmosphere absorbs most wavelengths of infrared radiation, astronomers try to get as high as they can with balloons, rockets, or satellites to get a clear view of the infrared universe.

One last detector reveals the remaining parallel universe and, in my mind, the most interesting one. Observing with this instrument, you do not see night at all. The sky is awash in light, the entire cosmos from one side to the other glowing a uniform white. The cosmic microwave background universe radiates more energy than all the galaxies together, accounting for as much as 99 percent of all the light energy in the universe. The sight is startling, even blinding, as if you are lost in a dense, bright fog in San Francisco. You know for a fact that buildings, autos, and people lie behind the fog, but its uniformity and pervasiveness hide them completely. A radiation fog blanketing the cosmos, the microwave universe occupies a wide band in the radio universe.

Radio waves range in length from a few millimeters to about 10 meters (from about ⅛ inch to about 30 feet), with those from about 8 millimeters to several meters easily traveling through the atmosphere without being reflected back into space by the ionosphere. The cosmic microwave radiation can be measured at wavelengths from 0.4 millimeter to about 700 millimeters (0.02 inch to about 28 inches). It was not coincidental that radio astronomers using a ground-based instrument to examine a 70-millimeter wavelength were the first to detect the cosmic background radiation.

Its most remarkable feature is the astonishing uniformity that perplexed its discoverers and makes it so difficult to measure. That, along with the fact that it is the most visible remnant of the very hot universe shortly after the Big Bang, made the cosmic microwave background radiation the most interesting thing in astronomy to me.

A RICHNESS OF EMBARRASSMENTS

The fact that no dipole anisotropy had been detected conclusively* in this uniform background radiation was a major embarrassment to Ray Weiss and other cosmic background radiation astronomers by the mid-1970s. Gradually becoming another embarrassment was the microwave background's startling uniformity—its utter lack of discernible landmarks. The dipole would have been one landmark, but it would be related only to the motion of the earth through the cosmic background. There would be no great cosmic significance to even convincing evidence that the dipole existed other than that theory was sound and instrumentation was capable of proving it so.

From a cosmic perspective, everybody realized that the cosmic background radiation had to appear uniformly smooth throughout the heavens. Penzias and Wilson already had determined this was true. But the radiation could not be absolutely smooth for the simple fact that the universe of today is not smooth. It is filled with unequal portions of matter and energy unequally distributed through space in the form of stars and galaxies and huge clusters of galaxies such as the Great Wall, discovered in 1989 by Margaret Geller and John Huchra of the Harvard-Smithsonian Observatory. From the day he first started speculating about it quantitatively while working with Robert Dicke at Princeton in 1964, Jim Peebles had recognized that the radiation, if it existed, would have had a major effect on the way galaxies had formed.

The major concern of cosmology for the following three decades was to figure out how the Big Bang, which apparently had deposited the cosmic background radiation evenly throughout space, led to the formation of stars, galaxies, and the immense galactic clusters and superclusters. Peebles realized that in the early universe—the first 300,000 years or so—the radiation dominated everything else. The ratio of energy in the form of photons to matter in the steaming cosmic soup of the infant universe was approximately the same as it is

---

* It had been measured correctly, it turns out, but Ray was not convinced.

today: as many as one billion photons to each particle of matter. But these photons were packed so tightly together during this period when the universe was its densest that their energy completely overwhelmed the ability of this fledgling baryonic matter to condense into atoms.

About 300,000 years after the Big Bang, scarcely a tick of the cosmic clock, the energy level of the photons that made up the radiation dropped as it attenuated—or became diluted as the wavelengths of the photons were stretched out—in the universe's expansion.[1] Matter could now begin condensing into hydrogen and helium atoms, the simplest kinds. Radiation, no longer the universe's dominant influence, and matter simply separated for good at this time, almost never to meet again; and gravity became the great shaper of the universe down to the present time.[2] Matter never again interacted much with the radiation left over from the Big Bang. Peebles and another theorist, Joseph Silk at Berkeley, reasoned that the radiation seen today might still contain clues to how matter was distributed throughout the universe at the moment matter and the radiation parted company. In effect, departing matter should have left an imprint on the radiation that should be apparent now. But in the 1960s such an imprint was not apparent, nor was it in the 1970s (or 1980s).

In 1967 Silk recommended that somebody map the temperature of the cosmic background radiation across the sky in every direction with sensitive instruments in order to obtain a snapshot of the universe at the moment matter separated from radiation, allowing galaxy formation to begin.[3] Minuscule fluctuations in the temperature of the cosmic background radiation should reveal where matter had been densest and hence where galaxies already had started to form.

Silk also pointed out that if the cosmic seeds of galaxies could not be found, cosmologists should begin looking around for another theory that could better explain the existence of the universe.[4]

BEAN COUNTING

Officials at NASA headquarters in Washington greatly underestimated the enthusiasm and proposal-writing ability of scientists who had an idea for an experiment in space. Announcements of opportunity No. 6 and No. 7 were widely distributed throughout the U.S. scientific community and abroad; it was expected that they would draw a number of ideas. Even so, NASA officials had anticipated few more than a score or two of responses. They had reasoned that there were only so many kinds of scientifically meaningful experiments that could fit aboard a satellite carried into orbit by a Scout or Delta rocket, but a flood of proposals poured into NASA headquarters in the fall of 1974—some 121 in all.

NASA officials immediately realized that the huge number would seriously strain their ability to give each proposal a serious and fair evaluation, especially since the responses covered such an extended array of topics. The proposals, sent in by groups from every part of the nation as well as from Europe and Japan, included earth sciences such as geology and oceanography as well as a number of ideas to study the magnetosphere, the ionosphere, the upper atmosphere, and the Van Allen belts. Other groups wanted to look farther out, proposing to study the Sun, the Moon, and the planets and their moons. The rest of the proposals covered most parallel universes in the astronomical spectrum: X-ray astronomy, ultraviolet astronomy, a flare survey for stars, and an ultraviolet orbiting observatory.

Our newly formed science team was proposing to create just the kind of map of the cosmic background radiation's temperature that Silk had suggested seven years earlier with the four instruments we wanted to send into orbit. Following our September 1974 meeting we somehow wrote a proposal. With our group scattered up and down the East Coast, this was fairly difficult in the days before fax machines and the Internet. We exchanged ideas by telephone, U.S. mail, and even in person. I traveled back and forth from New York City to Washington on Amtrak, waking early to catch the 7:30 train from Pennsylvania Station. Bob Silverberg usually picked me up at the station in Lanham, Maryland, and drove me over to Goddard for meetings with him and Mike Hauser.

Ray Weiss and I worked on the proposal together, I as chief writer and he as editor-in-chief. We also had begun meeting frequently with Goddard engineers and had begun depending heavily on their input about what was possible to do with scientific instruments in orbit. They filled us in on how NASA worked in terms of "costing" a project—that is, estimating its budget—and setting schedules. One of our biggest worries was about the technology we planned to use that did not yet exist. I had more faith in NASA engineers than did Ray, who was deeply worried that the soaring costs of creating this new technology would simply run us out of the competition. I talked with him one day in October by telephone about this problem.

"I don't see how you're going to keep this thing under $50 million and do all the things that we want to do," Ray said. "But if we don't send up all four instruments, there's no point in doing it at all."

Ray reminded me again of the need for multiple instrumentation in order to eliminate galactic noise by looking at the cosmic background radiation from cross perspectives. A quirk of NASA's system of cost accounting might work in our favor. "NASA doesn't keep honest books," said Ray, who was familiar with how the agency determined costs from his work on NASA science review committees. By this he alluded to the way NASA accountants were required to calculate the cost of an individual project: All the costs were not assigned directly to the project itself, but many cost overruns, extras, add-ons, and overhead costs were absorbed elsewhere in the NASA budget; in particular, if NASA decided to expend civil service manpower on a project, those salaries would not count against the project's budget.* This might save our expensive mission: If given the agency's blessing, it would become a part of the Explorer scientific program, which lay entirely within NASA and would not be subject to detailed congressional scrutiny. Its costs would be borne wholly by the NASA budget. But, of course, we had to convince NASA of our proposal's merits over more than a hundred competitors.

Ray and I put the final touches on the proposal, the breakdown of our copying machine contributing to the stress at the end. We finished it just before the deadline, October 28. Team members signed off on it, and the required twenty copies were driven down to NASA headquarters.

The proposal was incredibly ambitious. Looking back on it now, I am sure such a scientifically unified concept would not stand a chance in the cost-cutting, less confident climate of today. It called for a Cosmological Background Radiation Satellite that, we naively asserted, would answer all the remaining scientific questions about the cosmic background radiation. We would test the blackbody nature of the spectrum with a spectrometer that we stated was an improvement on the one I had designed for my ill-fated thesis. I thought we might be helped by the fact that, in the technically adept hands of David Woody, the instrument had by now flown successfully on a high-altitude balloon and produced some useful numbers.[5]

---

*Only in 1996 did federal agencies attempt full cost accounting, in which each project budget includes all direct and indirect costs.

We stated that our satellite would find the cosmic background radiation's missing anisotropy with both microwave and far infrared detectors. We also would measure the cosmic infrared background radiation at shorter wavelengths. The payload would be launched into orbit aboard a Delta rocket. There would be a helium cryostat containing three instruments in the center of the satellite, with four microwave receivers attached to the outer shell of the cryostat. A folding sunshield would protect the instruments from the Sun and the earth.

All in all, I thought, it was a very nice but thin document—just forty-nine pages plus three appendices discussing the possibility of putting liquid helium in space and listing the science team's biographies. The science was exciting and our instrument concepts were sound, although scantily detailed. Most important, perhaps, was that our team consisted of several real experts in the cosmic background radiation field (all with far more experience than I).

Estimating costs was a new adventure for me. In spite of our talks with engineers at Goddard and in private industry, none of us knew how to cost out our instruments which, our proposal blindly stated, would require $2.9 million to $4.85 million. Based on estimates from Ball Brothers Research Corporation in Boulder, Colorado, the cryostat would be another $2.5 million. Everything considered, the total cost of the satellite, cryostat, and instruments, along with two years of data analysis, would be between $28 million and $35 million. Looking back on these figures, I see that we were hopelessly innocent. On the data analysis alone, we were off by a factor of about 100, and even counting for a factor of 3 for inflation, we still missed the total cost of the data analysis by a factor of 30.

This naïveté was to come back to haunt us over and over as time went by.

INTERESTS CONFLICT

By now Nancy Boggess was working full-time at NASA headquarters. She immensely enjoyed the day-to-day bureaucratic tussle and the chance to meet some of science's leading lights, who would stop in from time to time at the NASA building on Independence Avenue. One of these was Lyman Spitzer, a grand old man of astronomy from

Princeton who was the first to suggest that the kind of nuclear fusion that made stars shine could one day be harnessed to produce electricity, an idea that had come to him while riding a ski chairlift at Aspen. Spitzer also was the first with the idea of putting a permanent optical telescope in orbit. Nancy also met Ray Weiss when he had first come to NASA seeking funds for his balloon experiments to measure the cosmic background radiation.

This was Nancy's first direct exposure to experimental work on the cosmic background radiation. She was so fascinated by what she learned from Ray that she took him up on his offer to visit Palestine, Texas, for the launch of one of his NASA-funded experiments. As I learned later, this was typical of Nancy. When something attracted her full interest, her natural enthusiasm simply took over and she would bear in on her new subject, bright blue eyes blazing. Nancy and her husband, Al, were avid birdwatchers, and it was their customary style to pack up their collection of binoculars and head off to the ends of the earth—Madagascar or a remote Aleutian Island—to track down a single species known to roost there. If they missed it one year, they thought nothing of going back again the next.

Nancy was among the officials at NASA headquarters overwhelmed by the flood of proposals generated by AOs No. 6 and No. 7. Her boss, Nancy Roman, set up an ad hoc committee for each discipline consisting primarily of outside scientists brought in to help judge the proposals. It was the agency's policy that every committee be chaired by a NASA employee. Nancy R. appointed Nancy B. to head the infrared committee, her area of greatest personal interest. This was a plum assignment. The prestigious National Academy of Sciences had recommended to NASA that the highest priority be given to a satellite-borne infrared experiment (the academy also had suggested that cosmic background experimentation could be done most effectively from balloons, a recommendation our science team did not know at the time).

NASA, moreover, already had opened negotiations with a research team in the Netherlands that had conceived and begun designing an infrared astronomical satellite to be called the Astronomical Netherlands Satellite. Apparently the Dutch group, headed by Reinder van Duinen, was in a hurry to be the first up with an infrared satellite, and wanted to keep moving. Scientists from the Netherlands had spent $1.4

million on their design and were working with Ball Brothers Research Corporation (now the Ball Aerospace Division) in Boulder to develop and build the liquid helium cryostat they would need and to design their telescope and detectors. The Dutch intended at that time to build the spacecraft themselves as well as run its communications and were negotiating with the British to help with data analysis.

Infrared astronomy was the hot subject among U.S. scientists, too, with twelve proposals—a tenth of the proposals NASA received— relating to the infrared universe in one way or another. All the big names in infrared astronomy came in: Russ Walker from the U.S. Air Force Geophysics Laboratory near Boston, who headed an impressive team of well-known infrared astronomers, including Gerry Neugebauer from Caltech, along with Frank Low and William Hoffmann from the University of Arizona; this one came with about sixty endorsements of major scientists, including Charles Townes. Also submitting proposals were Martin Harwit and Jim Houck from Cornell, and Al Harper at the University of Chicago. Mike Hauser, my colleague on the cosmic background radiation experiment, had submitted one from his new infrared group at Goddard Space Flight Center. There was another team from NASA's Ames Research center in Mountain View, California, headed by Ed Erickson, and a six-member group from the Jet Propulsion Laboratory in Pasadena—all asking NASA to let them be the team to put an infrared telescope into orbit aboard a satellite.

One of Nancy's problems was finding competent scientists outside NASA to review the infrared proposals, since virtually everyone of note in the infrared community was involved in submitting a proposal. She eventually had to seek committee members not particularly knowledgeable about infrared such as Vera Rubin, a top-notch optical astronomer at the Carnegie Institute in Washington. Nancy also had asked Ray Weiss. This presented her with her second problem. Since nobody knew what to do with the small trickle of cosmic background proposals that had come in, NASA officials decided to consider them along with the infrared proposals. This made perfect sense: scientifically, because the two occupied the same end of the astronomical spectrum, and practically, because so many of the problems designing and building instruments were the same. Ray agreed to join the committee. But what would she do with him when our Cosmological Background Radiation Satellite proposal came up? Quite obviously, he could not sit in judgment of his own proposal. Since the stakes were so high, both scientifically and economically, NASA could face a lawsuit. But would judging other competing proposals also be considered a conflict of interest?*

"NASA lawyers had looked at the AOs very carefully," Nancy recalled.[6] "After reviewing the situation regarding Ray, they finally gave their permission for him to sit on the infrared committee. But when his own proposal came in for discussion, he was to leave the room and not discuss it with the other committee members."

The committee met in early 1975 at NASA headquarters over a two-day period to review the proposals. Nancy chaired the meetings and set their agendas. As they began to discuss the proposals, committee members decided to assign each one a numerical ranking: No. 1, a well-thought-out, excellent proposal deserving full consideration; No. 2, a proposal with merit but needing more work; and No. 3, a proposal not worthy of consideration. No. 3 was obviously the kiss of death, but No. 2 was virtually the kiss of death, since the committee really would consider only proposals of real merit. The committee first took up Russ Walker's proposal. His group already was way out in front of everybody else in concept and design and had a huge list of endorsements from the community. The big Walker collaboration proposal was categorized No. 1, as were the Cornell and Goddard proposals.

Then the committee approved the idea on the spot, deciding to form a unified team from all three Category 1 infrared proposals.

The first job of this new infrared team was to decide about collaborating with the Dutch team, although there was not much doubt it was the right thing to do.

"This was not announced publicly at the time," said Nancy. "In fact, it was very hush-hush, and the committee members were asked not to discuss what they had done outside the meeting room." NASA wanted to make sure that *all* the proposals were being given a full, impartial, and democratic hearing. The National Academy of Sciences had already given a high recommendation that an infrared sky survey was a

---

*During this well-publicized review period, Nancy Boggess had other concerns too. A U.S. Senator called her, wanting to know whether NASA could predict the alignment of the planets in 2050 for a constituent's horoscope. Nancy politely advised him to contact the U.S. Naval Observatory in Washington, where such information was kept.

top priority, and it was not necessarily inconsistent with NASA policy to approve some Category 1 proposals for study before others. But the quick approval might give the impression that committee members were giving short shrift to the remaining proposals. In fact, they did read them carefully.

The committee was somewhat at a loss when it took up the Cosmological Background Radiation Satellite proposal. It was given a category No. 1, but nobody knew quite what to do with it. After much discussion comparing it with the infrared experiment, the committee reached a consensus—"more or less," according to Nancy's recollection—that a cosmic background radiation experiment would be unique. Moreover, it would not conflict or overlap with the infrared experiment that the committee had just approved. But what did this mean? Since it seemed almost unthinkable that this small subcommittee possibly could get *two* separate satellites, the committee decided that the spectrum experiment on the cosmic background proposal was the highest priority. After further discussion, the reviewers voted to recommend that the cosmic background people get together with the infrared people. It would then be determined whether it would be feasible simply to bolt the spectrum experiment piggyback-style onto the back of the infrared astronomical satellite.

This was now being called IRAS and was obviously, in the minds of the reviewers and agency officialdom, the basket into which all the NASA infrared eggs were going to be placed. Where did this leave our experiment? I had no idea, of course, not being privy to the NASA selection process. A few weeks after the reviewers met, I was invited down to give an oral presentation on our proposal. I had little time to prepare, and was quite nervous. This fear was enhanced further by a chance meeting with Luis Alvarez, the celebrated Nobel Prize winner from Berkeley who was there to talk about a proposal he was submitting. I was aware that members of his group had been working on the cosmic background radiation, but I had no idea that they had submitted a proposal to NASA. Alvarez's presence there was an unnerving surprise.

When it came my turn to enter the committee room, I walked in and gave my hastily prepared presentation. One of the reviewers asked me if it would be possible to attach our experiment onto the IRAS. I said I thought it would. "How much would you estimate the cost of such a hook-up?" the reviewer asked. I thought we might be able to build our detectors in the laboratory, then prior to launch bolt the instruments onto the IRAS satellite for the ride into space.

"Well, my thesis project cost about $100,000," I said. "I would imagine that it probably would cost somewhat more than that."

Everybody in the room laughed, including me.

✳ 10 ✳

Dueling Satellites

*If you want to be a scientist, you've got to have a daddy [or a mommy]
helping you along the way.*

—RAY WEISS

ON SEPTEMBER 10, 1975, I found myself seated on a transatlantic air-
liner en route to the Netherlands courtesy of Nancy Boggess.
During the flight I was preoccupied with drawing a new design for our
cosmic background radiation spectrum experiment that would allow it,
in shrunken form, to be lofted into space as part of the infrared astronomical satellite payload. We landed early in the morning. After
clearing customs at Schiphol Airport, I went straight to conference
rooms in the hotel and later at the Leiden Observatory and the University of Groningen to meet with the Dutch planners of IRAS. They had
arranged the session for everybody, Americans and English alike, who
conceivably could become involved with the project. There was a broad
agenda, but the sessions of most concern to me involved the feasibility
of placing additional experiments aboard the IRAS, our project, of
course, possibly being one of these piggyback riders.

A sequence of unpredictable events, at least to me, had sent me to
Amsterdam. Following my presentation to the review committee, I
received a letter from NASA appointing me to serve on the mission definition team for the IRAS. I learned later that this appointment was the
result of efforts by Nancy Boggess to keep our cosmic background radiation experiment alive at NASA headquarters despite a general lack of

interest among the powers that be, who were much more excited by IRAS. Because NASA had received so many proposals for infrared experiments, Nancy was able to persuade her superiors to create a U.S. team from competing groups. In the ensuing bloodbath, some big names like Martin Harwit dropped out. Among the survivors—including Frank Low from Arizona and Mike Hauser, who also was on our team—there were many conflicting ideas and a lot of heated words. Gerry Neugebauer from Caltech and Reinder van Duinen, head of the Dutch team, emerged as the scientific team leaders who would run the project.

The question of cooperation between the two national groups, with British scientists playing a somewhat smaller role, was very tricky. The good aspect of an international team was that NASA officials believed they might be able to undertake a challenging space project within the Explorer program budget. At the time it was believed that the U.S. contribution might be as little as $30 million, not including NASA launch expenses. U.S. scientists also liked the idea that an international agreement would be harder for NASA to cancel as difficulties arose. Moreover, the Dutch had a good head start and were willing to let us join them in order to gain similar benefits. They also were excited about the prospect of expanding their space industrial base in the Netherlands.

The flip side of a joint American-Dutch space venture, of course, was that divided responsibilities would be difficult to manage, especially across an ocean. There would be many chiefs, all wanting control. Worse, from the perspective of some U.S. scientists, was that many of the scientific ideas of the Dutch were simply horrifying. In order to keep costs down, the Dutch had made many compromises. The most serious was reducing the amount of data that would be sent down by the satellite. They had done this by putting smart star recognition equipment into the flight electronics that would send back answers to limited questions anticipated years before launch, rather than simply sending back all possible raw data. What if the sky were much more complicated than anyone could anticipate? It would later turn out that dust clouds, for instance, were just as interesting as stars, but nobody would have seen them without the raw data.

Having no prior experience with this particular Dutch group, the U.S. scientists agonized over such cost-cutting measures, fearing the worst: that they would be unable to convince the Dutch of anything substantive, much less anything relating to such fundamental design questions as the data-analysis system, which, after all, was the heart of the experiment.[1] It took some time for the careful thought process and professional expertise of the Dutch to be fully recognized by the aggressive American team. During the meeting in Amsterdam I suggested that our cosmic background radiation experiment could be tucked in above the secondary mirror of the infrared telescope, where it would have its own little peephole from which to look out into the cosmos. A few others there had similar ideas. Richard Jennings of University College in London was interested in an experiment to study cosmic background radiation anisotropy using a little device down in the focal plane of the telescope.

Another British scientist, Peter Clegg, suggested sending up a polarizing Michelson interferometer, while Hans Olthof, a Dutch scientist, wanted to use a Fabry-Perot spectrometer to examine the interstellar molecular hydrogen line at 28 microns.* Nothing was resolved at the meeting. On the way home I stopped off in London, where I visited Clegg and Jennings to talk over the experiments and their technology. The big question for us was whether my little flared horn design was the proper one, or whether a more traditional approach using lenses, mirrors, and baffles would be sufficient.

At another meeting in Amsterdam the next year, I again presented my spectrometer concept to the joint U.S.-Dutch team, this time as part of a more formal agenda. Nancy Boggess, who was shepherding IRAS through NASA and feared the Dutch might walk away with the experiment, was there along with Ray Weiss. By now it was becoming evident that the IRAS project was going over budget and would be risky enough even with no additional experiments. During my talk, I observed that we were making some fairly radical assumptions about what we could accomplish in the tiny volume of the satellite that might be available to us. Nancy was trying to keep our project alive and aboard IRAS, the reason she had made sure I attended the conference. But Ray was not optimistic. One day he took me to a restaurant to talk over the precarious state of our project.

"Forget it," Ray said of my design for mounting a cosmic background experiment behind the infrared telescope on the satellite. "It's never going

---

* Many of these ideas had to be delayed for more than twenty years. Several instruments similar to those discussed in Amsterdam in 1975 were launched on the Infrared Space Observatory in 1995 by the European Space Agency.

to work, even if you do manage to get it on the IRAS satellite. You're not going to be able to do the science you want to do with it. It's going to be compromised so badly that there's no point in doing it at all." Ray was a realist as well as a visionary. I had to admit that he probably was right.

At the meeting it was becoming clear that a majority of team members had little interest in the scientific results we hoped to obtain. I had the distinct feeling that both the U.S. and the Dutch infrared scientists already had decided that their own science was going to be far more exciting than our spectrum experiment. They seemed to think our instrument was immature and that, perhaps, I was too, meaning that I was not in a very good position to carry the project off. I already had started getting the same message from back home.

At one meeting at Goddard, Gerry Neugebauer said he thought I was too young. He may have been right. I was eager and relatively quick to learn, but I had little experience in the political maneuvering or in the fundamentals of space technology necessary to carry off a big project in Earth orbit. At a meeting of the Society of Photo-Optical Instrumentation Engineers in Reston, Virginia, it also had become evident to me that there were practical reasons for NASA to push IRAS ahead of our project. Infrared technology could be used for looking down through clouds day or night, so it could easily observe missiles after launch as they rose up through the atmosphere.

Enormous defense contracts were already in place for developing infrared technology for the spy business. The contractors had developed a huge array of new technology and were rich with defense money. This was both a benefit and a problem for IRAS. In the mid-1970s the detectors being planned for the infrared satellite relied on classified technology then available. But tapping into this availability was difficult, and NASA was forced to spend much time and money negotiating for permission to use the technology in an international space project where it was feared infrared techniques could fall into the wrong hands. Fear of the USSR made infrared science possible and then threatened to take it away.

GOING IT ALONE

Within a year Nancy Roman and other space science officials at NASA decided to drop further efforts to place our spectrometer, or any other supplementary experiments, aboard IRAS. This was, unfortunately, a good management decision, although had I known of it at the time I certainly would have wondered about my future. The final IRAS carried a simple camera, which was an array of detectors packed closely together in the focal plane, and a very basic, low-resolution spectrometer with a grating to disperse light before it entered the detectors. They weighed little and were extremely simple in concept, in fact much less complex than our spectrometer.

Even so, the IRAS project was unusually difficult to complete and eventually ran far behind schedule and over budget, taking seven years to launch with a number of small disasters along the way. Fortunately, it was worth the trouble: IRAS was able to see several hundred thousand new stars and tens of thousands of external galaxies, make beautiful maps of the sky, and even turn up a few stars that appeared to have the debris of protoplanets circling the stars in disks around them. Back at headquarters, Nancy Boggess had reluctantly agreed that it would be virtually impossible to place our instrument aboard IRAS and have both experiments accomplish their goals.

"But we've still got to do the cosmic background experiment," she told her boss, Nancy Roman. "It was a Category 1, too." Nancy Boggess believed a cosmic background radiation experiment could provide some important clues about the origin of the cosmos. "I want to know as much about the universe as I can in my lifetime," she said. "I am absolutely sure a cosmic background experiment in space will tell us a lot about the infancy and evolution of the universe."[2]

Aided and abetted by Ray Weiss, she decided to push strongly for a cosmic background radiation experiment aboard its own satellite. Nancy was willing to use brinksmanship within NASA and work almost any angle she thought would convince the powers-that-be that a cosmic background instrument in space would be a plum for NASA. Still, it would not be an easy feat. An internal memo circulating at NASA headquarters listed the following priorities for space science: (1) IRAS, (2) an X-ray satellite, (3) an Ultraviolet Explorer, (4) a cosmic background radiation experiment. Moreover, members of the original review team feared the expense and difficulties associated with putting two big dewars into orbit aboard separate satellites. There was little enthusiasm for a cosmic background radiation experiment, since it was already apparent that NASA would have to pump a lot of money from its precious Explorer program into IRAS* The reviewers were right; our project would be expensive and difficult.

Since this program was a line item in the NASA budget, money for a project that had concluded could be shifted over easily to another project without congressional approval. But when would funding for IRAS stop? It was huge and hugely expensive. What was the point of starting a second project dependent on the same money pool with no end to the first one in sight? Complicating matters further was the fact that by now Nancy was having her own problems at headquarters. She had worked hard to learn the technical side of the IRAS project and was so successful that Gerry Neugebauer had invited her to become a member of the IRAS science team.

"Nancy, you're one of us," Neugebauer told her. But this invitation did not go over well with some NASA officials, who criticized Nancy for delving too deeply into individual projects and for discussing their merits with outside scientists.

"Your job is not to do your job too well, but to do many jobs," one of her superiors warned her. Nancy Roman, her immediate superior, was opposed to approving a cosmic background radiation experiment at this time, believing that it didn't have a chance to make it at the next level of competition. Nancy Boggess strongly believed that Roman was mistaken, and that her lack of flexibility in looking at the overall picture of space science could doom the cosmic background radiation project. Nancy Boggess decided to take control of the bleak situation herself.

During a period when Roman was away from headquarters, Nancy Boggess went with Mike Hauser to see Franklin Martin, the head of NASA's astrophysical sciences division. She showed him our proposal along with two others for a cosmic background radiation experiment that had also received a Category 1 rating from the review committee,

---

*Despite the lack of enthusiasm at NASA headquarters for a study of the cosmic background, a study by the agency's Astronomy Missions Board as early as 1969 identified it as "one of the most challenging problems in modern astrophysics," and concluded that "as a fundamental cosmological phenomenon, the microwave background has a high priority for study from space."[3] In January 1976, NASA's Outlook for Space study group identified the "nature of the Universe" as a major theme for future space objectives, listing the measurement of the spectrum and isotropy of the cosmic background as a key question that could be addressed by techniques in space.[4]

but they had been placed on hold while Nancy tried unsuccessfully to have our experiment placed on board the IRAS. Nancy and Mike talked to Martin about the importance of these experiments for the science of cosmology.[5]

"These experiments can have a major effect on our understanding of the way the universe developed," Nancy told Martin. "Right now there's a big gap in this understanding, which these experiments can fill. Besides, if they work the way they're supposed to, it will be a big plum for NASA."

Martin was convinced, and so was Noel Hinners, NASA's chief of space science, whom Nancy also had worked hard to persuade. They agreed to approve seed money for the study of the feasibility of sending a cosmic background radiation experiment into space aboard its own satellite. Nancy was elated. On March 1, 1976, she wrote a letter of congratulations to Ray Weiss, Mike Hauser, Dave Wilkinson, and me—from our original team (NASA dropped the other three members)— instructing us to join forces with the two scientists from other groups to form a mission definition team. Noel Hinners would be the responsible manager at NASA headquarters, while Nancy would be the program scientist. We were told to determine what experiments to put aboard a cosmic background satellite and how best to get it into orbit.

When Nancy Roman returned to headquarters, she was taken aback by what Nancy Boggess had done. "Are you sure about this cosmic background experiment?" Roman asked. "It doesn't have a chance of making it." Nancy Roman felt that Nancy Boggess probably was stringing those proposing it along for naught. Nancy Boggess was not sure it would make it either. But for the time being her gamble had paid off.

THE COALITION

A new team soon began to emerge phoenixlike from the ashes of Nancy's unsuccessful effort to place our spectrum experiment aboard the IRAS satellite. The other cosmic background radiation proposals had come in from a group at Berkeley and another team at the Jet Propulsion Laboratory in Pasadena. The Berkeley group was headed by Luis Alvarez, whom I had run into at the proposal presentation at NASA headquarters the year before. Samuel Gulkis, whom I did not know at the time, was the single member of the JPL group who survived NASA's initial selection process. NASA headquarters apparently believed strongly that it had the right to create a forced marriage of disparate groups who had submitted competing proposals, then add or subtract people from the final team as it wished. NASA had done this with IRAS. When our turn came, the agency dropped Bob Silverberg, Dirk Muehlner, and Pat Thaddeus.

Gulkis, along with Michael Janssen and six other colleagues at JPL, had submitted a very fine proposal to NASA calling for a switched radiometer to be placed in orbit. It was based on the original concept pioneered by Robert Dicke in the 1940s, and Gulkis proposed to use the instrument to look for anisotropy in the cosmic background radiation. He was a radio astronomer who had received his Ph.D. in physics from the University of Florida in 1965. At the time Penzias and Wilson discovered the cosmic background radiation, Gulkis was working with the big antenna at Arecibo, Puerto Rico, to study weak radio sources. After a year's stint as an assistant professor at Cornell, Sam moved to JPL in 1968. He helped develop the masers and low sidelobe microwave horn antennas that were used for communicating with planetary missions such as Voyager, Mariner, and Pioneer.

Working with JPL's 64-meter antenna at the Goldstone Tracking Station in the California Desert, Sam had tried to find the roughness in the cosmic background radiation that Jim Peebles had first predicted.[6] He did not find it, but he did set new limits for ground-based observations of the cosmic background radiation. The Gulkis group's proposal to NASA was remarkable for its insight into what could be learned from looking for large-scale anisotropy in the cosmic microwave background. These included the possibility that the universe is expanding at different rates in different directions, or that gravitational waves of great amplitude are distorting local space-time. They noted the Sunyaev-Zeldovich effect, in which hot electrons in clusters of galaxies could distort the cosmic microwave background radiation. They even included a table of predicted amplitudes for this effect.

Sam and his group also analyzed the dipole anisotropy, which had not yet been discovered (predicting a level within 10 percent of that which was eventually found), and speculated about the relationship of the universe's so-called missing matter with the cosmic background radiation (which is still not well understood). They picked a small instrument they believed would stand the best chance of being selected for NASA's Explorer program. Their payload had a launch weight of only 264 pounds (120 kilograms) and would cost only a few million dollars—perfect for a small Scout rocket.

The radiometer design was a traditional one that would use a Dicke switch to compare data from the sky with an on-board reference blackbody emitter and absorber in a waveguide. This had the virtue of keeping data analysis simple, but the disadvantage of lost sensitivity since the instrument would be looking at the sky only half the time. It would take measurements of the whole sky at three frequencies between 18 and 55 gigahertz (1 GHz is a billion cycles per second) selected to eliminate the effects of foreground galactic emissions from the hot electrons of the Sunyaev-Zeldovich effect. Knowing that the receivers then available were not very sensitive, Gulkis and his colleagues planned to make maps with an angular resolution of only 15 degrees and a sensitivity of 0.3 mK per pixel.* In retrospect, such an instrument would have detected virtually nothing. But, of course, nobody knew this at the time.

A solar orbit far from Earth would be the best location for their instrument, the JPL group believed. But, hoping to keep costs low and their chances for acceptance by NASA high, they sought only a simple equatorial orbit even though they recognized that it would be difficult to avoid problems caused by the Sun and the earth shining onto the apparatus, which would affect both its temperature and the resulting data.[7]

"We didn't even know at the time that a polar orbit was a possibility," Gulkis recalled.[8] "And there was a culture of conservatism at JPL that didn't encourage trying something that new and radical. Maybe the people at Goddard already knew that the military had achieved polar orbits. But we didn't. So we went with the equatorial orbit even though we were aware of all the problems it could cause."

A big plus for Sam and his colleagues was that they already had helped place microwave radiometers into orbit for a weather satellite called Nimbus. The radiometers were capable of looking down through clouds or darkness at the earth's surface to map its temperature. The

---

*A *pixel* is the smallest element of an image, like the dots in a halftone photograph. The sensitivity would be a part in 10,000, which was very good for those days.

detectors could be designed with a spectrometer to determine the emissions of particular atmospheric constituents, such as water vapor. Another plus was that Sam's group had developed a new kind of microwave antenna, called a dual-mode conical corrugated horn, which is now the standard in the field. Its design reduced to a very low level the instrument's response to undesirable sources outside the field of view.

The group from the Lawrence Berkeley Laboratory,* a facility of the University of California at Berkeley, proposed to use this same kind of horn antenna for their cosmic background radiation experiment. I had met the team's leader, Luis Alvarez, a few times at Berkeley and respected his mind and his fangs. Luie, as he was called, was born in 1912, the son of Dr. Walter Alvarez of the Mayo Clinic in Minnesota. He had worked during World War II at the radar lab at MIT, where he invented an instrument ground approach landing system for aircraft that played an important role in the war, and later in civilian aviation. While there, typically working alone since he would not be led, he also invented continuously adjustable focal-length spectacles for the presbyopia of middle age that a patient could simply dial in for him- or herself. (There is no indication that this good idea ever reached the marketplace.) He had worked on the Manhattan Project and won the 1968 Nobel Prize for major contributions to high-energy particle physics at Lawrence Berkeley. He had the vision to recognize that the cosmic background radiation was important and had started members of his group working to take new measurements of it. His interests were diverse. He examined the interior of an Egyptian pyramid using the only particles that could go through an enormous amount of rock: cosmic rays. Alvarez found no surprises inside, but his unique exploratory method put him in the news.

Several years later he became famous for a theory he proposed with his son Walter, a geologist, that the dinosaurs had become extinct after a giant comet or asteroid meteor struck Earth 65 million years

---

*The Lawrence Berkeley Laboratory, a nonmilitary facility, should not be confused with the Lawrence Livermore National Laboratory, a nuclear weapons research facility in Livermore, California. Both facilities were named after the same man, Ernest O. Lawrence, who helped develop particle accelerators and founded the Berkeley laboratory, and both were funded by the federal Atomic Energy Commission.

ago. The idea originally was considered wacky, but is now accepted in many paleontology circles because strong evidence of a strange and violent event has emerged. A layer of rock was found coinciding with the end of the dinosaur era; within had been found enhanced iridium. This could have occurred if an iridium-rich meteorite struck the earth, filling the sky for years afterward with clouds of iridium-laced dirt and rock particles. Luie was innovative in many areas, even inventing an image-stabilization optical system for binoculars. Its mechanism was passive, making it much less complex than the gyroscopic packages used to stabilize images in video cameras and other optical systems.

In physics circles Luie was considered brilliant, creative, and terrifying. He was known for unmercifully battering his postdocs and other scientists during seminars and oral presentations. I attended a meeting in Berkeley once when Luie tore into another scientist, a physicist named Mike Lampton (who later became an astronaut). I decided right away that Luie was not a person with whom I would enjoy working, although this may have been an overreaction on my part. I was well acquainted with one member of Luie's group, Rich Muller. His wife, Rosie, was a Swarthmore alumna, so we had met at Berkeley and gotten on friendly terms even though we never actually worked together.

Rich was bright and inventive and had received a number of honors. He was treated well by Alvarez. Rich later became well known among physics insiders for disproving a claim that gravity waves had been detected. An inevitability of general relativity, gravity waves should occur when a big event such as the explosion of a star shakes the fabric of space-time. They might be detectable when they strike Earth. Joseph Weber maintained he had found evidence of gravity waves with a bar antenna detector he had invented at the University of Maryland. Rich persuaded Weber to let him analyze one of his data tapes from the detector and found no evidence of a gravity-wave event—contrary to Weber's claim.

Another of Alvarez's team was a young physicist about my age named George Smoot. He had received his undergraduate and doctoral degrees from MIT, then come to Berkeley as a postdoc to work with Alvarez. Although we had overlapped at Berkeley, I had not met him because he worked in the Lawrence Berkeley Laboratory while I had been a graduate student in the university itself. George was tall with red hair and a beard, and was always on the lookout for something new to try. Working with Alvarez, he had become interested in cosmic rays and was extremely involved with a NASA plan to place a huge cosmic-ray detector into orbit. The instrument would include a large supercon-ducting magnet in a liquid helium dewar. George was awarded a patent for his invention of a method to contain the liquid helium. Helium-4, the normal isotope, is *diamagnetic,* meaning that it is repelled by a magnet. In space this repellent force would be strong enough to keep the liquid from rushing out the vent. Great effort and much money were expended on studying the project for many years, but NASA never built it.

In the 1960s Alvarez persuaded NASA to begin funding a high-altitude balloon project at Lawrence Berkeley. Joining Alvarez's group in 1970, George had become heavily involved in a number of balloon-borne experiments. One of the earliest sought to discover evidence of cosmic antimatter. After briefly thinking they had found such evidence, George and Luie decided they had not after all. On six more balloon flights George and his colleagues, including Rich Muller later, began searching for dipole anisotropy in the cosmic background radiation. Later they used high-altitude U-2 planes to provide a more stable and predictable platform for their instruments.[9]

Alvarez encouraged intragroup rivalry, believing it brought the quickest scientific results. Tension always existed among group members, especially between Rich Muller and George Smoot. Rich was an idea person while George was the hands-on type, which served to exacerbate the stress between them when scientific theories conflicted with technical possibilities.

"It really was quite infantile," recalled Philip M. Lubin, a graduate student in Alvarez's group at the time (and later a member of our sci-entific team).[10] "By the time they started working with the U-2, things had become very bad." During one argument Muller and Smoot exchanged threats of physical violence. Alvarez did little to intervene because he was working on his dinosaur theory with his son and had little time to give to the search for dipole anisotropy. Eventually George managed to have Rich ejected unceremoniously from the U-2 project, even though Alvarez clearly favored him.

The Lawrence Berkeley proposal to NASA listed Alvarez as prin-cipal investigator and eight co-investigators, including Smoot and

Muller.[11] Smoot was really the leader of the proposal, and Alvarez was listed as P.I. because Smoot did not yet have a high enough position in the organization. The proposal called for a small Scout-launched instrument weighing a few hundred pounds, similar to Sam Gulkis's design. But the Berkeley team identified the preferable polar orbit, although one at low altitude to save launch weight. They hoped to achieve a sensitivity of 0.2 mK and an angular resolution of 5 degrees, which was much more ambitious than what the JPL team had in mind. Alvarez's group stated that they planned to use some very good amplifiers they believed could do the job. They planned to use just two frequencies, although they noted that more would certainly be better (and more expensive).

Their scientific rationale was similar to that of Sam Gulkis's group. A notable addition was that the Alvarez team also wanted to try to track down evidence that the universe could be rotating as a whole. If it were, there ought to be a recognizable pattern in the brightness of the cosmic background radiation with two poles and an equator showing up. Ernst Mach, a nineteenth-century Austrian physicist, had developed a hypothesis that inertial reference frames are not innate in a body, but are governed by their relation to all the other mass in the universe.*

An object would experience no inertial forces when it is at rest or in uniform motion with respect to the center of mass of the entire universe. According to the Berkeley proposal, if the universe were seen to rotate, it would challenge Mach's fundamental idea: How could the universe rotate relative to itself? As they learned later (along with everybody else), the rotation aspect of Mach's principle is a direct result of general relativity, meaning that it would be possible to test Einstein's theory by studying the behavior of the matter and space-time contained within a rotating shell of distant matter such as the entire universe.

Unlike the Gulkis group's design, the Berkeley satellite would not spin but would position one face always toward the Sun. This was easy and cheap, no doubt a choice required by what they felt would be a

---

*Mach is better known for devising the Mach number, the ratio between the velocity of an object and the speed of sound in the medium in which the object is traveling. Hence an airplane with a velocity of Mach 3.0 travels at three times the speed of sound as measured in prevailing atmospheric conditions.

minuscule budget for the mission.* Although more sensitive than the JPL version, the Berkeley radiometers were much more complicated. They planned to use parametric amplifiers and dual antennas, with a Dicke switch flipping the input data back and forth 100 times a second between the two antennas. The amplifiers used a well-known technology, but making them work at high frequencies was high art. Because of the amplifiers' notorious instability, it would have been necessary to spin the satellite, we now know in retrospect.

The Berkeley proposal sought $200,000 from NASA for further study and named George Smoot, who was spearheading the proposal, as its contact person. Luie Alvarez was the headliner for the time being, the person the Berkeley group hoped would catch NASA's eye and win approval for them. They succeeded. At NASA headquarters, Nancy Boggess and the review committee had been duly impressed. Luie was among the first people Nancy called as she began forming a new NASA-certified science team. She asked him whether he would be interested in becoming a member of the new NASA team to plan a cosmic background radiation mission.

By no means, he replied. Alvarez told her that the project was becoming far too large and cumbersome as far as he was concerned. He preferred a relatively quick, contained experiment that could be done by a small, fast-acting group. Besides, Alvarez pointed out, he was too old at sixty-five to begin a project that could take years or even decades to bring to fruition. He might not live to see its conclusion.† He noted that there were other qualified people listed on the title page of his proposal. One of them could take his place. Nancy and Luie discussed several different members of the Berkeley group. She asked him for his recommendation.

"Any one of these you pick will be fine with me," Alvarez told her.[12]

---

*From our present-day perspective, we know that it would have been very difficult to obtain good data without spinning the satellite in order to swing the horn antennas rapidly across the sky. Most microwave radiometers simply are not stable enough to maintain a steady reading for the entire time it takes a satellite to make a complete orbit of Earth.

†Unfortunately, Alvarez proved prescient on this point. He died in 1988, the year before the COBE was launched.

Nancy narrowed her choices down to Rich Muller or George Smoot, both of whom she already knew because, through her, NASA had helped fund their U-2 aircraft experiment searching for cosmic background radiation anisotropy. She knew George was especially eager to have the job and was looking for something new and different to try. Moreover, he was younger than Rich and might remain more enthusiastic over what promised to be the very long haul of the project. She picked up the phone and called him.

NASA, working through Nancy Boggess, had now created a six-member team—Gulkis, Hauser, Mather, Smoot, Weiss, and Wilkinson—that would lead the cosmic background radiation experiment down through the agency's own bureaucratic gauntlet into the coming decades and eventually, we all hoped, into space.

✶ II ✶

Mathematical Mythology

Give me matter and I will construct a world out of it.

—IMMANUEL KANT

GODDARD SPACE FLIGHT CENTER brings to mind a junior college campus or a 1950s military base. Bike paths wander pleasantly through partially forested terrain and alongside a pond where geese occasionally breed, the facility's twenty-nine spartan government-issue buildings mostly of beige brick the only detraction from the bucolic setting. Named for the rocket pioneer Robert Hutchings Goddard, the center was created in May 1959, at the height of cold war fears about Soviet dominance in space. Its 1,100 acres were carved hastily out of the U.S. Beltsville Agricultural Research Center in the gently rolling Maryland countryside about 10 miles northeast of Washington.

Today Goddard has nearly 12,000 employees—about 3,700 civil servants and the rest contract workers—and an annual budget of more than $1 billion. It operates a satellite facility near Wallops Island on the Atlantic Coast of the Delmarva peninsula (where Al Boggess had helped launch small suborbital scientific rockets in the 1960s). Goddard also oversees the small, academic Goddard Institute for Space Studies in New York City, where I worked as a postdoc with Pat Thaddeus. Goddard always has been the most scientifically oriented of NASA's nine U.S. field centers, with more Ph.D.s than any other. Central to Goddard's elaborate scientific enterprise are the Space Science

Directorate and the Earth Sciences Directorate. They manage large laboratories that undertake studies of the atmosphere and oceans, Earth and space physics, high-energy astrophysics, astronomy, and planetary sciences.

Among these is a Laboratory for High Energy Astrophysics that employs, among others, astronomers and theorists who work at trying to comprehend the origin, structure, and fate of the universe. In one of those quirks of accidental timing that in retrospect appears destined, I came to work as a full-time civil servant at the laboratory on June 1, 1976. I had been offered a job several months earlier at Goddard, but did not know it. The official letter was never sent as the result of a misunderstanding, and I was on the verge of taking a job at Bell Labs in New Jersey. Mike Hauser, whom Frank McDonald had hired to start an infrared group the year before, had arranged for Goddard to hire me and recruited me vigorously. When he heard I was about to go to work for Bell Labs, he called. Mike was passionate in his belief that our proposal for a cosmic background radiation experiment in space eventually would become a real project. As we talked, the mystery of why I had not responded to the unsent offer was solved.

"It's worth coming to Goddard on the chance it will fly," he said. The fact that he was in a good position to know was encouraging. I accepted the job, drove down from New York, and moved into Mike's home for a couple of weeks. The first thing I bought was a piano.

One night Mike and I drove to Hamburger Hamlet on upper Wisconsin Avenue in Washington with our dates. We talked about our proposal over dinner, mulling over a name for the project. A good name was important. It would help sell our project to NASA, just as punchy names like *Big Bang* and *black hole,* a term recently coined by John Wheeler of Princeton, had helped sell those theories to the public. The four of us tossed around a few ideas. Cosmic Radiation Satellite. Cosmic Background Satellite. Microwave Background Pioneer. Cosmic Background Radiation Explorer. None of the acronyms worked. "MBP" or "CBRS" were eminently forgettable. "Try something else instead of the first initials," somebody said. The thought cleared our minds.

"What about Cosmic Background Explorer?" it was suggested. (Nobody remembers who first uttered these words.)

It clicked. We could call it COBE, pronouncing it "coby." The name fit our proposal perfectly, said what we thought it should say and, we hoped, was just catchy enough to get some attention.

Within two weeks of starting my new job I received a letter from Noel Hinners at NASA headquarters with exciting news: Approval had been granted to study the possibility of an entirely distinct cosmic background radiation experiment that would be launched into orbit aboard its own satellite. Hinners suggested I drop off the IRAS team in the best interests of the new project now under study. Ray Weiss and I both quickly submitted our resignations, although Mike Hauser decided to remain on the IRAS team. He had been the principal investigator on the Goddard infrared satellite proposal two years earlier at the time of the announcements of opportunity. Since then he had earned the respect of the merged IRAS team, even though he was not yet one of the big names in infrared astronomy.

A week later I got more encouraging news, this time from George Pieper, Goddard's Director of Space Sciences. He was naming me study Scientist, the Goddard employee who officially would be responsible for the scientific conduct of the project during its study period. What this really meant, I learned during the ensuing weeks, was that I would have to get down to business. Not only would I begin working with an engineering manager to identify the many difficult technical problems that lay ahead, but I would begin purchasing equipment and working with government contractors. In reality, most of the purchasing and contract negotiations would be handled by our study manager at Goddard, a person still to be selected. But I would be involved, meaning that soon I would be getting my feet wet in what was to me the slick-bottomed pond of the business world.

COSMIC COURTSHIP

As our new team was being formed, we decided to send out a COBE newsletter. "Unique results crucial to cosmology will be obtained," I grandly wrote in the first issue. The use of the indicative rather than the subjunctive mood was an apparent sign of my optimism that the COBE would one day fly in spite of the many "ifs" still strung out ahead of us. We expected to find and study in detail, I continued, "anisotropies and deviations from a thermal spectrum, cold galactic dust, hot galactic dust, integrated light from stars, zodiacal dust emission, galactic synchrotron radiation, molecular and atomic emission lines from large gas clouds in the galaxy, and integrated emission from local and remote extragalactic sources."[1]

We distributed our newsletter with its ambitious catalog of hopes and expectations far and wide through the cosmology and astrophysics communities. It turned out to be the first and last COBE newsletter we published. But it not only stimulated excitement about our project; it also brought in a steady stream of new ideas. Among the recipients was Steven Weinberg, who would win the Nobel Prize two years later for his role in developing the theory that electromagnetism and the weak nuclear force that causes radioactive decay in some atomic nuclei were manifestations of the same phenomenon. At the time Weinberg received the newsletter, he was working on a book aimed at making the scientific ideas behind the Big Bang theory accessible to the public. He thought our plans worth mentioning (although enclosed within parentheses): "As final corrections were being made in this book I received a *Cosmic Background Explorer Satellite Newsletter #1* from John Mather of NASA. It announces the appointment of a team of six scientists, under Rainer Weiss of M.I.T., to study the possible measurements of the infrared and microwave radiation backgrounds from space. Bon voyage."[2]

His book *The First Three Minutes: A Modern View of the Origin of the Universe* was published the next year, and was an instant popular hit (and deservedly remains in print today). Everybody on the team considered our brief notice a minor coup, believing Weinberg's kind words could do nothing but help COBE's cause. Whenever I or other members of our group stopped by a bookstore, we bought a copy or two, then made sure the proper individuals in the NASA bureaucracy received a book with the appropriate page marked. I don't know whether our distribution of *The First Three Minutes* helped our project as far as officials at headquarters were concerned, but I am certain it did not hurt.

Weinberg only recently had brought his prodigious talent for formulating theories in particle physics to the scientific search for the universe's origin. His book inspired not only the reading public but the next generation of physicists and astronomers as well. *The First Three Minutes* also symbolized the coming of age of cosmology as both theoretical and experimental science. This exciting development was brought about by the marriage of astrophysics and particle physics, which Weinberg helped achieve. The courtship went back to the 1930s, when members of the two branches of the physics community concerned with the most fundamental affairs of the cosmos—atoms and stars—took to the floor for a slow dance that eventually would lead to matrimony. As often is the case, a simple question initiated the courtship—in this case, Where did the energy for the stars come from?

A British astronomical theorist, James Jeans, constructed a model for the structure of a star at the turn of the twentieth century: a large mass of gas drawn together by its own gravity with an extremely hot and dense core. Jeans believed that the energy radiating outward from the seething center of the gas made the star hot and bright. Intrigued, William Thompson, Baron Kelvin, one of the most imposing intellects of the day whose main interest was heat and electricity, estimated that if Jeans's scenario were correct, a mid-sized star such as our Sun would burn out in approximately ten million years.

This was damning news. Geologists already had found rocks and fossils they were certain were far older than ten million years, which quite rightly implied to them that the solar system must be much older still. Moreover, Lord Kelvin's calculation was wholly inconsistent with Charles Darwin's new theory of evolution: Ten million years was simply insufficient time for the thousands of biological mutations in millions of organisms the theory required.

Could Darwin, the geologists, Jeans, and Lord Kelvin all be right? Astronomers began thinking that a source of energy other than the gravity holding it together must be powering the Sun (and the other stars too). But if not gravity, what? Arthur Stanley Eddington had been among the first to grasp the revolutionary way of thinking brought about by Einstein's equation, $E = mc^2$, when others hesitated. He recognized its implications for the stellar fires: The conversion of matter into energy could be the only process capable of producing their astonishing energy. A possible mechanism, Eddington suggested, was the conversion of hydrogen into helium. If you had 4 kilograms of hydrogen and were able to take the protons that made up their nuclei and reconfigure them to form helium nuclei instead, you would end up with 3.97 kilograms of helium. The difference in mass, 0.03 kilograms representing the binding energy of the nuclear constituents, would appear as the light and heat of stars. In his masterful 1926 book, *The Internal Constitution of the Stars,* Eddington christened his process "the transmutation of elements." Now it is known as fusion.*

Still, nobody could explain the full process. In the 1930s European physics was thrown into turmoil by the rise of the Nazis in Germany and Stalin in the USSR. Many of the best and brightest scientists fled for their lives, among them Edward Teller from Hungary and George Gamow from the Soviet Union. Both worked on the stellar fusion problem at the University of Göttingen under the famous Max Born before emigrating to the United States in 1933 and 1934, respectively. Reunited, Teller and Gamow organized a small conference in Washington to solve the fusion problem once and for all. One of the attendees was another refugee from the Nazis, Hans Bethe, who was teaching physics at Cornell. He was so quick with numbers that he often left other physicists and even mathematicians far behind with his remarkable problem-solving talent. It took Bethe just six weeks to work out the details of the nuclear fusion processes of both the Sun and bigger, hotter stars.†

Bethe's explanation of the nuclear fusion powering the stellar furnaces that light up the universe was the first step down the aisle in astronomy's marriage with particle physics. He demonstrated how four hydrogen nuclei could fuse into a helium-4 nucleus (a couple of electrons also are used up and some neutrinos are released, but this was not known then). Bethe's process accounted for the energy source of the Sun and most of the other main sequence stars. During the next decade he refined his explanation of stellar fusion, which eventually led to a Nobel Prize in 1967. Gamow, Alpher, Herman, and a few others began working out the mechanism by which certain particles had been created during the hot, early moments of the universe.

---

* Eddington was criticized by other physicists who believed that the interiors of stars were not hot enough to produce nuclear fusion. Eddington believed fusion could occur at temperatures far lower than most physicists believed possible. He invited his critics to "go and find a hotter place."

† After sending his calculations to *Physical Review* for publication, Bethe learned of a $500 prize being offered by the New York Academy of Sciences for the best unpublished paper on stellar energy production. He asked the journal's editor to return his paper, entered the contest, and won. He used some of the prize money to help his mother emigrate from Germany.

A few years later Fred Hoyle turned his skill in quantum mechanics on a rare class of stars out of the main sequence, red giants. In 1954 he hypothesized that these stars were fired by an unusual kind of nuclear reaction in which three helium nuclei combined to form a carbon-12 nucleus but only at a certain quantum energy level. Three years later Hoyle's friend and collaborator, William A. Fowler, used an experimental apparatus at Caltech to find to his astonishment a carbon-12 energy level exactly where Hoyle said it should be—the first time an astronomical prediction had been tested and confirmed in the laboratory.

The long courtship between subatomic physics and cosmology would seem to have ended in 1965 when Penzias and Wilson discovered the cosmic background radiation. Astronomical prediction and electromagnetic technology were now seemingly bound together forever.

In reality, the cosmic nuptials were not fully consummated. Too many problems remained for particle theorists to work out. One was antimatter, which was deeply perplexing. The idea that for every particle there was an antiparticle of equal mass but with the opposite electrical charge had been predicted by the British theorist Paul A. M. Dirac in the 1930s. Carl Anderson, a Caltech physicist, confirmed the prediction when he found a positron, or positive electron, in a cloud chamber four years later. When a positron and electron encountered each other, Anderson discovered, the two simply vanished in a mutual annihilation, the only remainder a pair of gamma rays.* Anderson's work, which had the beneficial side effect of winning Nobel Prizes for both Dirac and him, was confirmed when bigger accelerators found that an antiparticle existed for every particle.

But where was all this antimatter? The universe should consist of equal amounts of it and matter, since there is no known reason why nature would prefer one over the other. Yet there was no sign of antimatter on Earth. And from studies of cosmic rays, which bore a few antiprotons from deep intergalactic space, there apparently was no more than 1 part per 10 billion throughout the entire cosmos. In 1967

---

*In 1956 Harold Furth, then a young physicist at the University of California at Berkeley and later director of the fusion project at Princeton, published a poem called "The Perils of Modern Living" in *The New Yorker* (August 5, 1956). Furth imagined a meeting between Dr. Edward Teller and Dr. Edward Anti-Teller. "Their right hands clasped," wrote Furth, "and the rest was gamma rays."

Andrei Sakharov, the humanist and pioneer of the Soviet hydrogen bomb, suggested a possible explanation for the ratio between anti-matter and matter in today's universe. During the first 10^{-6} second after the Big Bang, the universe could have contained a slight excess of matter over antimatter, estimated at about 1 part in 10 billion. Within the next fragmentary second, matter and antimatter collided in a colossal annihilation. Left behind were only radiation and the remnant matter. The universe of today—its stars, its galaxies, its black holes, its single known life-bearing planet—consists of this residual material.

Sakharov went further. He calculated that the current background temperature could be accounted for in a universe that started out with nearly equal numbers of particles and antiparticles. The energy of annihilation of the antiparticles would have warmed up the cosmic background radiation the amount observed today. This provided a hypothetical answer to an old conundrum: Why is the temperature of the cosmic background radiation what it is?

Other theorists worked to refine the Big Bang model. One was Stephen Hawking, the remarkable Cambridge University scientist who is confined to a wheelchair because of amyotrophic lateral sclerosis. He collaborated with Roger Penrose, an outstanding Oxford mathematician and member of a celebrated British scientific family, to describe the first instant of the Big Bang.[3] By the late 1960s nobody had been able to describe the state of the universe at that moment. Most theorists believed this was simply a theoretical abyss, a moment beyond the beyond: a point where space and time would disappear in a singularity where the field equations of general relativity broke down in a mathematical thicket of infinities and meaningless zeros. Such a singularity was the metaphorical end of the road back to the instant of creation.

Penrose was familiar with singularities. In a mathematical tour de force a few years earlier, he had demonstrated how a star collapsing under its own gravitational mass would physically end up at such a point. A star was one thing, Penrose knew. The entire universe was another matter. One of the problems theoreticians had faced unsuccessfully was how to figure out what would happen to all the newly created particles during the extreme heat and density of the universe's first instant. A group of Russian theorists had proposed in 1963 a hypothesis calling for quickly alternating contracting and expanding phases that would keep particles from striking one another.

Penrose and Hawking decided on a simpler approach: Why not look at the way points in space-time might be causally related? Such thinking permitted them to avoid analyzing individual particles as everybody else was doing and concentrate on the equations of general relativity in order to look at the properties of space-time itself during the instant of the Big Bang. They found that there could have been a singularity at the beginning of time. The surprise, though, was that their calculations appeared to *require* a singularity.[4]

SCIENTIFIC CREATION

The Big Bang is the most scientific of the creation mythologies.* It plays a theme similar to more ancient ones: a time before time, a darkness before light, something uncreated before something created. In ancient Polynesia the sea god Tangaroa lived alone above a vast expanse of water. He threw down a stone. It became land. He sent a bird to plant a vine, which rotted. As it decomposed the maggots became people. The Greeks pictured a void called Chaos. Gaea, mother of creation, emerged from the darkness of Chaos and founded the dynastic order of the god who would rule from Olympus. In ancient Egypt the divine he-she serpent Atum lived in the black waters of Nun, a primordial ocean in which lay the seeds of all creation. Lonely, Atum masturbated to create the first creatures, the male Shu and the female Tefnut, giving them *ka,* the vital essence.

These mythologies are haunting projections of the human mind trying to find order in the bleak, random, and lonely universe in which it found itself. The creation epic of twentieth-century physics varied little from these others, differing mainly in the degree to which it quantified its causal chain: The universe, beginning in an extremely hot and dense fireball still not well understood, expanded. Matter, consisting of quarks and radiation, cooled. Protons and neutrons coalesced from the quarks and, within a few minutes, nuclei of helium, deuterium, and lithium began forming from the protons and neutrons.

Still cooling and expanding some 300,000 years later, these nuclei began combining with electrons to form electrically neutral atoms. The

---

*Most scientists probably would argue that it is the *only* scientific version of creation since, in principle, it is falsifiable by experimental test and by logical deduction.

creation of such atoms allowed matter and radiation to separate for the first time and begin evolving independently in what is known as the decoupling era. Free of pressure from the radiation, the new matter, influenced mainly by gravity, eventually was drawn together into stars and galaxies. Stars created not only more chemical elements but also a large amount of radiation that should now exist as the yet-to-be-seen cosmic infrared background radiation.

Another kind of remnant radiation, the cosmic microwave background radiation, of course, had been studied closely for more than a decade by the end of the 1970s. In our current universe it is faint and cool after attenuating during the universe's expansion since the decoupling area. It still remains the universe's dominant form of radiant energy. It has the properties of a nearly perfect blackbody and a temperature of approximately 2.7° Kelvin. By measuring these two fundamental properties of the cosmic microwave radiation—its spectrum and its uniformity—across the sky, we are in a very real sense taking a snapshot of conditions in the universe 300,000 years ago.

Observations had shown that the radiation's spectrum was so close to that of a blackbody source that it seemed quite clear that the universe was nearly in a state of perfect thermal equilibrium at the moment of decoupling. Even minuscule variations from a blackbody spectrum would indicate that highly energetic processes had disrupted the thermal equilibrium before the decoupling era. A departure from a blackbody spectrum also could imply that large amounts of matter had altered the radiation after decoupling; this deeply concerned theorists unable to account for it in their emerging picture of the evolution of the universe.

Careful measurements of the cosmic background radiation should, we thought, provide further clues about the universe's evolution. Slight differences in temperature in different directions across the sky could have been produced by an unequal distribution of matter and energy during the decoupling phase. These small variations in the distribution of matter must have served as the cosmic seeds of current galaxies and galactic clusters, if the theorists were right.

THE PROMISED LAND

The history of the universe as depicted by the theorists during the years following the discovery of the cosmic microwave background radiation was a lovely and ingenious moving picture running backward to the dawn of time. But the script was incomplete, still containing big gaps and inconsistencies that theory alone might never be able to fill in. If, for instance, gravity had been the only force driving the evolution of the universe on the cosmic scale, it would be difficult if not impossible to reconcile an absolutely smooth temperature of the microwave background radiation across the sky with the plainly visible lumpiness of the cosmos today.

Searches for anisotropy in the background radiation had been extensive and acutely sensitive. But by the early 1980s every result was negative. On angular scales greater than a few arc minutes, scientists could state with certainty only that the cosmic background radiation was smooth to within a few hundred parts per million. Theorists believed they could explain in some detail the earliest moments of the universe. But they still were unable to describe fully what had happened from the instant of the Big Bang until the formation of the most distant quasars being observed. This was a period of approximately one billion years, obviously a significant epoch in the universe's history.

Another problem had been largely ignored because, quite simply, it could not be solved. In 1969 Charles Misner, a young astronomer who had worked under Robert Dicke at Princeton, began wondering why the cosmic background radiation was so utterly smooth. Like Magellan sailing a ship toward terra incognita, astronomers figuratively journeyed by means of their telescopes on faint waves of radiation out toward the edge, or so-called horizon, of the universe. Looking out at the very limits of observability, they discovered that the cosmic background radiation flowed in from all directions at exactly the same temperature: about 2.7° Kelvin. Widely separated regions of the universe could have arrived at the same temperature only by having been in contact at some time in the past, Misner thought. But could this have happened?

According to the Big Bang picture, the horizon was about fifteen billion light-years away in any direction. The background radiation from one end of the universe would take fifteen billion years to reach Earth, just like the radiation from the opposite horizon. Misner reasoned that the two regions of space must be separated by at least thirty billion light-years. But the universe was supposed to be only about fifteen billion years old. How could these vastly remote regions of the universe ever have exchanged heat and achieved identical temperatures?* There simply was no way for a signal moving at the speed of light, the maximum velocity obtainable, to have traveled between any two points separated by more than 2° of angle at the time of decoupling.[5] Looking back earlier in the universe's history, Misner realized that the problem was worse than he feared. When the universe was only about 100,000 years old, not even a cosmic toddler, the presently observed regions of space would have been separated by approximately 10 million light years. This was not a great distance in today's universe. But at $t = 100,000$ years, this distance was 100 times greater than light possibly could have traveled since the Big Bang.

Like a rock in your shoe when you're on a long walk, Misner's horizon problem nagged and irritated cosmologists for more than a decade. There was no apparent answer. The problem was set aside until 1979. That year a young theorist in elementary particle physics, Alan Guth, started wondering about an unrelated problem. Guth, who had received his undergraduate and graduate degrees at MIT, was a postdoc at the Stanford Linear Accelerator working out theoretical problems of particle physics related to the Big Bang. A few of the current theories were adept at predicting the number and kind of subatomic particles seen throughout the universe today. Most of these, though, had an inconvenient consequence. They inevitably predicted that a particle known as a magnetic monopole should exist today in staggeringly large numbers: approximately 10^{80}, about the same as the number of protons in the universe. Several experimenters were actively searching for these monopoles, but with little apparent success. "I was wondering if there were any assumptions that could be changed to make theory compatible with the fact the universe did not seem to be swimming in magnetic monopoles," Guth recalled.[6]

Short and cheerful, he had a shaggy mop of dark hair that made him look like a bespectacled Beatle.

One evening at home he began wondering what would have happened had the universe, instead of expanding at an even rate, cooled rapidly right after the Big Bang. In a moment of epiphany after his

---

*The horizon problem caused no difficulty in Fred Hoyle's steady-state universe. Since such a universe had existed forever, there would have been adequate time for any two regions of the universe, no matter their distance today, to have exchanged thermal energy.

family had gone to bed, he suddenly realized that this might have led to an extremely rapid, though short-lived, period of expansion. What could have caused this, though? Guth knew that, according to particle theories, the universe had experienced a rapid change called a phase transition as it cooled in the instant after the Big Bang. When water is chilled very rapidly, it can remain liquid far below its freezing point of 0° Celsius for a moment; then it freezes all at once. This was what must have happened right after the Big Bang, Guth thought. As the universe cooled, the instantaneous false vacuum created by supercooling had driven the rapid expansion.

Later Guth christened this "inflation." He calculated it would have begun about 10^{-35} second after the Big Bang, when the hyperdense conditions would have created the false vacuum. According to general relativity, a kind of antigravitational force should have pushed matter apart instead of drawing it together, occurring so quickly it could be grasped only mathematically. Within the vanishingly small span of 10^{-32} second, the antigravitational repulsion would have made the universe expand in size by an astonishing factor of 10^{50}. Smaller than the nucleus of an atom, the universe would have inflated to a diameter of 10 centimeters —roughly the equivalent of a grain of sand growing to the size of our universe in the same instant. (See figure 3.)

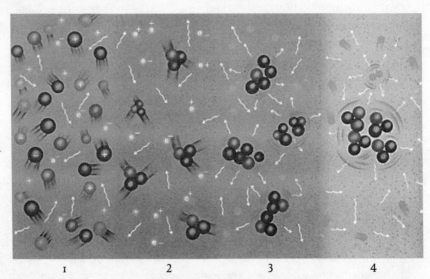

| 1 | 2 | 3 | 4 |

Fig. 3. HISTORY OF THE UNIVERSE.
Eight phases of expanding universe. *Left to right:* (1.) Quarks and exotic elementary particles, time = 0.001 second. (2.) Photons and neutrons. (3.) Antimatter annihilation.

The inflationary period ended when its inherent instability caught up with it, according to Guth. The universe then reverted to the more leisurely pace of expansion of the standard Big Bang model. If it had actually occurred, though, the stunning instant of inflation would have been the most critical period in the universe's entire history. Guth and others who, hearing the news, jumped on the inflation bandwagon worked on the equations of the inflationary epoch for months. Gradually, it became apparent that inflation could resolve many troubling anomalies. The monopole conundrum was quickly dispatched, for the inflationary model predicted that only a single monopole, perhaps one or two more, would now exist in the universe.*

More important, the horizon problem no longer existed. Before inflation began at 10^{-35} second, every region of the universe that we observe today would have been in contact with every other. In other words, the universe had evolved from an exceedingly small space that had been in

---

*In 1982 a Stanford experimentalist named Blas Cabrera startled the physics world by announcing that he might have found a magnetic monopole in an instrument built for that purpose. The result has never been duplicated. Had Cabrera by some vanishingly small chance captured the single monopole predicted by inflation? It seemed unlikely, considering how much effort and money were expended in future searches. Cabrera never claimed a detection, but felt obliged to reveal the data.

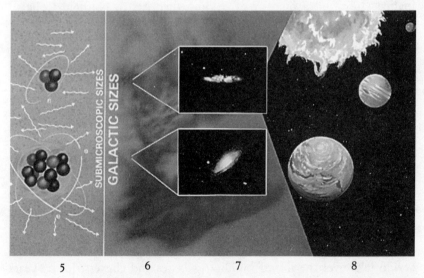

5   6   7   8

(4.) Formation of helium nuclei. Time = 3 minutes. (5.) Formation of atoms. Time = 300,000 years. (6.) Formation of galaxies, time = 200 million years. (7.) Galaxies. (8.) Solar system at present, time = 7 billion to 15 billion years.

COLLEGE OF THE SEQUOIAS
LIBRARY

thermal equilibrium prior to the time inflation began. The cosmic background radiation was smooth, according to the inflationary model, because its photons issued from a single region of thermal equilibrium.

Inflation theory also handily dealt with another difficulty. For years astronomers had been puzzled by the so-called flatness of the universe; this was the all but unbelievable balance of the universe between runaway expansion and gravitational collapse. Some scientists, including Robert Dicke, believed this already should have occurred. This flatness problem was reduced to a simple exercise in geometry by inflation. Whatever the curvature of space before the inflationary period, it would have been flattened during the rapid expansion in a way similar to an inflating balloon whose surface becomes flatter and flatter as the balloon enlarges.[7]

For theorists, inflation was a universal cure-all, and it made its creator a celebrity in cosmology circles. In a single stroke Guth seemingly had created a new vision of the Big Bang that was not only revolutionary but also simple and sublime. The inflationary model so dominated astrophysics that it might have won Guth a Nobel Prize had the Swedish Academy of Science been accustomed to giving out the physics award for theoretical work in cosmology. Among those deeply impressed with the inflationary model was Stephen Hawking. He invited Guth and a number of leading theorists to Nuffield College, Oxford, in the summer of 1982 for a series of seminars called the Nuffield Workshop on the Very Early Universe.

The Nuffield group focused on the idea that the universe before inflation was far smaller than an atomic nucleus. The laws of quantum mechanics must have governed. In such an infinitesimally minuscule space—one that would carry along in its expansion every aspect of the universe as we know it—quantum particles would flicker in and out of existence in a fluctuating energy field. Like waves upon the surface of the sea, these quantum fluctuations would have peaks and troughs. The Nuffield group theorized that the peaks produced by these quantum fluctuations could produce density variations. An instant later the inflationary expansion would draw these variations out sufficiently to become the seeds of galactic structures. Eventually gravity would take over the job of building galaxies and clusters. In the years following the Nuffield sessions, dozens of theoreticians proposed scores of ideas dependent on the basic inflationary premise. Each one seemed to modify, refine, supplement, or supplant Guth's original.

"The idea that the universe began with inflation offers a simple solution to many problems," said Stephen Hawking.[8] "Inflation accounts for the fine-tuning of the universe. This allowed the universe to expand as it has without collapsing back on itself like a black hole. It accounted for the fact that matter was not spread too thin for galaxies to form."

Hawking, like many other theorists, believed theory had arrived at the gate of the promised land by the early 1980s. "Is the End in Sight for Theoretical Physics?" Hawking entitled a lecture in 1980, meaning that theoretical physics soon would have all the answers to all the big questions, "say, by the end of the century," Hawking speculated.[9]

Many of my fellow experimental cosmologists were not convinced. In fact, some of us believed that inflation was a hollow theory, making virtually no new predictions that could be tested either by experimentation or observation. The inflation theories did seem to explain away a number of troubling characteristics of the universe: its horizon problem, its flatness problem, its clustered galaxies, and so forth, all accounting for the theories' enormous popularity. The reason for inflation's extraordinary explanatory power, however, was simple. The theory had been created *a priori* to explain phenomena that had already been observed and not the other way around. This meant the theory could never be disproved.

Questions remained that theory alone could not solve: How close to a blackbody is the spectrum of the cosmic background radiation? What if the spectrum were not a blackbody after all? How had the universe evolved during its first billion years? Where were the angular anisotropies in cosmic background radiation that seemed necessary if the Big Bang model were correct? What about the predicted cosmic infrared background that had never been seen? As the 1980s progressed, more questions would arise in the form of new astronomical observations than theorists could answer. The answers had defied the best efforts of many fine astronomers for years.

My fellow gumshoe scientists and I believed only a set of instruments such as we were planning for the COBE spacecraft could provide answers to the fundamental questions about the origin of the universe. But this would require making some of the most difficult measurements in science.

PART III

First Light

✳ 12 ✳

Day One

Grown-ups love figures. When you tell them you have made a new friend, they never ask you any questions about essential matters. They never say to you, "What does his voice sound like? What game does he love best? Does he collect butterflies?" Instead, they demand: "How old is he?"

—ANTOINE DE SAINT-EXUPÉRY

ON JUNE 28, 1976, I walked across the Goddard campus feeling buoyant. I was enjoying my new job enormously, along with the new colleagues who came with it. In less than a month our COBE project had started to pick up steam, making Mike Hauser look like a fortune-teller. I didn't know how much he had already done to arrange the future. We still had a lot of work ahead of us to get our project into space. But at the very least we were in the running, while a number of other fine proposals were not. My colleagues on the new NASA-created science team had all arrived in town for our first meeting. I knew two of them only from their proposals and was looking forward to meeting them in person as I walked into the conference room in Building 2. That the meeting was being held there was fortunate for us. Goddard had lost out to the Jet Propulsion Laboratory on an earlier bid to engineer and construct the IRAS satellite. Upper-echelon officials at Goddard were delighted that another big science project might come their way.

Everybody arrived: the four members of my original group who had survived NASA's selection process—Mike Hauser, Ray Weiss,

Dave Wilkinson, and I—along with Franklin D. Martin and Nancy Boggess of NASA headquarters. I knew Frank and Nancy only slightly from our earlier presentations and phone calls. Frank, head of NASA's astrophysics division, would serve as our program manager, while Nancy would be program scientist, both oversight jobs required by headquarters. Ron Muller from Goddard's Preliminary Systems Design Group was also there. He was our study manager and would lead us through the design and contract maze.

The two newcomers at the table, at least to me, were Sam Gulkis and George Smoot, who had both flown in from California. Somewhat stocky and of medium height, Sam was mild-mannered and gentlemanly. He was a radio astronomer specializing in the solar system at the Jet Propulsion Laboratory, a NASA laboratory run by Caltech, and had an air of quiet professionalism. George was bespectacled like me, about my age (a year older), and nearly as tall. He had lived in a number of places, including Alaska, while growing up and, though obviously quite intelligent, seemed somewhat withdrawn. George eventually would become the most controversial member of the team.

After introductions, Nancy took charge: NASA would give us a one-year study period for our combined cosmic background radiation proposals, now officially called COBE. At the end of fiscal year 1977 NASA would decide whether to proceed to the next level, the so-called Phase B mission definition. "You'll have to have a good scientific and engineering study and a good cost analysis to show that COBE deserves a ride into space," Nancy said.[1] Even if NASA selected our mission over the others still in the running, it would not be launched for at least four years after NASA headquarters gave its blessing to COBE to begin its official development phase.

Realistically, we were only in a queue for a place at the end of the line. We would not be able to start building our satellite until all funds deemed necessary for the IRAS satellite had been spent. COBE would not fly for five years at the earliest, most likely aboard a Delta rocket. In the event a Delta were not available, a smaller Scout-class rocket or even the space shuttle were possibilities. The latter was a discouraging prospect, since the first space shuttle was still under construction and years from its first orbital flight. The shuttle's big advantage was that it could lift a satellite weighing several tons. But it would be years before a launch facility at Vandenberg Air Force Base would be readied for launching shuttles into the polar orbit we wanted.*

"Unfortunately, COBE doesn't have very high priority at headquarters," Nancy said in apparent reference to the internal memo placing a cosmic background radiation experiment No. 4 on a list of four. "But the experiment is strong scientifically. And it presents NASA with a terrific opportunity to make a real contribution to fundamental knowledge."

This was sobering news. George and Sam thought our proposal might stand a better chance with just a single experiment, a souped-up version of the anisotropy instrument their groups had submitted independently. It could fit on a small, inexpensive Scout rocket, they pointed out. Ray Weiss refused to let the spacecraft define the experiments.

"What's the point of sending an entire spacecraft up with one instrument?" he asked. "What kind of science are you going to do with that? You may as well use a balloon."

George and Sam were overruled. We would push for the four instruments our original group had been planning all along: the spectrometer that was an advanced variation of my thesis experiment, a radiometer to look for anisotropy in the far infrared universe, a diffuse infrared background experiment to measure cosmic infrared background from the earliest galaxies, and the differential microwave radiometer to look for anisotropies in the cosmic background radiation itself. We voted to seek a budget of $265,000 to get started: $70,000 for a Ball Brothers study of the instruments and the large cryostat that would be at the center of our satellite, with lesser amounts going to team members for their individual studies.

We had to pick somebody to head our team, the man who would run the scientific show. Ray Weiss was the senior scientist there and had already become our scientific conscience. We elected him chairman

---

* No space shuttle had been launched into polar orbit by the mid-1990s. A facility for such a launch was being prepared at Vandenberg Air Force Base in the early 1980s and was nearly complete when the *Challenger* exploded in January 1986. Since then, plans for a polar-orbital shuttle mission have for all intents and purposes been abandoned because of safety concerns relating to the Vandenberg launch facility, and because the Air Force switched to using expendable rockets for its spy satellites.

unanimously. When we met again six weeks later—two days after my thirtieth birthday on August 7, 1976—Frank Martin brought along an unwelcome surprise from headquarters: news that we would have to submit a preliminary report months earlier than we had expected, in January for a NASA peer review panel that would go over all the Explorer proposals and establish a new priority list. "We expect to make a go–no go decision on the Explorer proposals then," Martin said.[2] The priority list would in reality be a cut list, it was apparent.

The competition at the peer review sessions, Martin revealed, would include a proposal for studying transient X-ray sources, another for looking at sources of X-ray and extreme ultraviolet radiation, two more ultraviolet proposals, a Solar Mesospheric Explorer, another satellite to be called the Equion for studying the ionosphere, and two others for auroral experiments. The IRAS satellite already had been approved and was scheduled next year for a "Hard Start"—meaning actual construction could begin—along with another scientific satellite, an Electrodynamics Explorer that would study electromagnetic waves and particles in the very high atmosphere. Most of these proposals were much further along than we were.

"The problem is simple," said Martin. "We've got more studies under way than we can ever fund."

We wondered if we really were still in the competition. The IRAS satellite sat at the head of the line. Everybody we had talked with at headquarters believed that an X-ray scientific spacecraft was next, with one of the ultraviolet proposals lined up after that. Moreover, some people in the astrophysics community had started grumbling that our cosmic background radiation spacecraft would not even be pursuing established astronomy but merely conducting physics experiments in space.

THE GANG OF SIX

Where did that leave COBE? Our new team dispersed from the summer meetings at Goddard with a heavy work agenda but without knowing its fate. As we got busy on our report, I realized that everything that had happened up to now—the AOs, our original team and its slim proposal, my awkward first presentation, the broken-down copying machine, all the sweat and worry—had been merely preamble. Now we had very little time in which to rethink and refine our individual goals into a single project that would convince tough-minded, dollar-conscious bureaucrats at NASA headquarters of its spaceworthiness.

A major factor in our favor, it was apparent, was that we planned to use the same kind of cryogenics system that IRAS already had begun developing.[3] A second, though equally expensive to build, presumably could avoid design and development costs. But we were going to have to speed up and streamline our proposal. We would need to concentrate on what we might be able to achieve scientifically in terms of technology available, such as then state-of-the-art microwave mixers and relatively modest infrared detectors. We had wanted to use more exotic alternatives like maser receivers, which had yet to be designed and built. This more exciting technology would have to wait until after January. We had to get something down on paper now.

Almost immediately we started talking with people in industry. I flew up to Boston to visit Honeywell and Block Engineering, companies with experts on commercial Fourier spectrometers. I also stopped in at the Air Force Cambridge Research Laboratory (later renamed the Air Force Geophysics Laboratory and now called the Phillips Laboratory Geophysics Directorate) at Hanscom Field outside Boston. A. T. Stair, one of the leading lights in Fourier spectroscopy, briefed me on the Air Force's latest unclassified technology. At the lab I also looked up Thomas L. Murdock, one of the top Air Force scientists working on infrared measurements in space. Tom had grown up near Goddard in Silver Spring, Maryland, then graduated from the University of North Carolina, where he had been a member of the Reserve Officers Training Corps, and gone on to the University of Minnesota to earn his Ph.D. in physics. In 1972, as a young lieutenant in the Air Force, he started working as an astrophysicist at the Air Force lab where he now worked as a civilian scientist.

Tom was project scientist on an Air Force experiment collecting the first infrared data from zodiacal dust near the Sun. The Air Force was interested in zodiacal dust because it wanted to detect Soviet missiles shortly after launch and the dust would severely limit their detectors' sensitivity. I was impressed with Tom's expertise in infrared astronomy, much of which, unfortunately, was classified, and filed his name away. He had practical experience with the way infrared instruments worked in space, and also knew where to look in industry for expert help. He had helped convince his superiors that it was essential for national security to acquire data about the infrared sky if the Air Force planned to use infrared detectors to look for the launch of Soviet missiles.

"The Department of Defense was on a fast track to get space surveillance up into orbit, so they started pumping tens of millions of dollars into near-infrared detectors," Tom recalled.[4] Later he argued with the Air Force to publish for the astronomy community at large the information they obtained at such expense, but he had little luck. In the coming months I and other COBE team members sought his expertise frequently, although we could not ask him about classified information.

In the early days, most of the Air Force's data remained secret. However, an 8½-by-11-inch map mysteriously fell into the hands of a group of infrared scientists at the University of Arizona. The map consisted of random spots, each with two numbers alongside it. Knowing the wavelengths that would interest the Pentagon, the Arizona astronomers were able to decode it and its coordinates as an Air Force infrared sky map. The astronomers immediately spotted what they recognized as several significant errors, notably a sign error from the data-processing software.

In 1975 the Air Force completed a comprehensive catalog covering 79 percent of the infrared sky based on data collected by suborbital sounding rockets fired from White Sands and the Woomera rocket base in Australia. The Pentagon wanted a map of the sky at 4, 11, and 20 micron wavelengths, which approximated that of objects at or near room temperature—that is, you, me, airplanes, and rockets. The theory was that if a good map at those wavelengths were available, the Air Force would not attempt to shoot galaxies, stars, and other celestial sources of infrared radiation out of the sky in the event of a national emergency.

Learning of this problem and others, a committee of top infrared astronomers, some with security clearances, including Frank Low from the University of Arizona and Mike Hauser, was dispatched to Boston to check out the Air Force survey. The infrared community thought the Air Force might have additional data that could help them better plan the instruments for the IRAS spacecraft. And, it was hoped, academic scientists could help set Air Force data right. While examining the data, both classified and nonclassified, the scientists learned that the

Air Force observers had found approximately 650 new celestial sources of infrared radiation at the 11 micron wavelength.[5] But Hauser, Low, and the other astronomers could confirm only a low percentage of these sources from other known observations.

Studying and comparing the data, the astronomers considered that the Air Force observers had made several mistakes: Their observational techniques had no mechanism for self-confirmation, meaning that just one signal in a single detector could meet all their criteria for identifying a new source. There was no way to account for several observations of the same source. Air Force computer programs, designed for completeness rather than reliability, were less scrupulous in weeding out suspicious signals than an experienced human might have been. They had been less than successful in eliminating instrument noise and hence avoiding false detections of celestial infrared sources. The Air Force instruments carried no shielding to eliminate particle radiation that could also provoke false signals.[6]

Its own heavy security had hurt the Air Force work. Military scientists had scarcely any contact with the larger astronomical community and therefore were unaware of techniques that would have helped them avoid cataloging erroneous signals. Moreover, since their real purpose was national security, they were not trying to solve the same problems that astronomers would undertake, and they didn't have a budget for a lot of complicated software. The IRAS mission had to do a much better job in charting the infrared sky or tens of millions of dollars would be wasted. "The IRAS instrument, mission profile, and data processing can be designed to overcome the problems of the AFCRL survey," the report concluded.[7] Remarkably, IRAS scientists already had anticipated every problem encountered by the military scientists and had taken appropriate measures to eliminate the errors from their own project. Subsequently, a second Air Force study was released with much higher confirmation rates.

Thanks to Mike Hauser, COBE scientists also took a serious look at the Air Force's problems. Obviously, we wanted to do everything we could to head off those kinds of difficulties. We kept the Air Force study in mind as we worked on our report for NASA, whose compilation was more difficult and time-consuming than I had anticipated. Writing it became only slightly more manageable when we dropped Ray Weiss's pet instrument, the radiometer that was to look for infrared anisotropy. We all realized that the instrument was simply too difficult to design, build, and operate properly. Ray did not object. Ron Muller wrote the part of the report dealing with the technical aspects of the spacecraft, while I struggled to make the science section cogent as we wrote and edited in sections submitted by other members of the science team.

A few key decisions had to be made. Noel Hinners at headquarters had asked the science team to respond to a new Announcement of Opportunity for a Spacelab mission. We were being asked in essence: Can COBE science be done from relatively low shuttle altitudes? The team discussed the idea at length. At just 100 miles up there still would be some atmosphere that undoubtedly would condense on our apparatus. Moreover, the shuttle might emit radiation, and its immediate environment would be filled with evaporating waste materials floating around in space. We concluded it was hopeless, and wrote Hinners declining to respond to the AO. Another question was whether to seek help from scientists outside the United States. Mike, Ray, and I all had been involved with IRAS and were aware of the problems of working with overseas scientists. We were worried about transatlantic work, fearing we would be spending more time on airplanes than planning COBE. In the end, too many cooks did not spoil the IRAS broth. The mission was a great success despite persistent arguments among team members.

"You're asking for nothing but trouble," Ray warned us. Without giving it much more thought, we let the idea of seeking international cooperation die by default. In retrospect, I think we might have benefited from more contact with overseas scientists in Europe or Japan.

Another question was which members of the science team were going to be in charge of each instrument. Ray, of course, had been elected chairman and we had dropped his instrument. It was starting to look like Mike Hauser would be in charge of the Diffuse Infrared Background Experiment (DIRBE) and I would head the Far Infrared Absolute Spectrophotometer (FIRAS) work. That left the Differential Microwave Radiometer, which would look for anisotropies in the cosmic background radiation and was the instrument of greatest interest to three members of the team: Sam Gulkis, George Smoot, and Dave Wilkinson.

COBE SCIENCE TEAM.
The COBE science working group. *Left to right:* Edward S. Cheng (Deputy Project Scientist), David T. Wilkinson, Richard A. Shafer (Deputy PI, FIRAS), Thomas L. Murdock, Stephen S. Meyer, Charles L. Bennett (Deputy PI, DMR), Nancy W. Boggess (Deputy Project Scientist, front), Michael A. Janssen *(behind her)*, Robert F. Silverberg, Samuel F. Gulkis *(front)*, John C. Mather (Project Scientist, FIRAS PI. *behind him*), S. Harvey Mosely, Jr., Philip M. Lubin, Edward L. Wright (Data Team Leader), Michael G. Hauser (DIRBE PI), George F. Smoot (DMR PI), Rainer Weiss (SWG Chairman), and Thomas Kelsall (Deputy PI, DIRBE). *Not present:* Eli Dwek.

Dave, the team's senior cosmic background radiation experimenter who had been in from the start more than a decade earlier, could have had the job for the asking; he was NASA's choice. But he preferred smaller-scale, university-style experiments and was not eager to manage an endless project. Sam and George both wanted the job. But Sam, a hands-on experimentalist, would take it only on the condition that the instrument could be built at the Jet Propulsion Laboratory where he worked. Although highly experienced at constructing instruments for spacecraft, JPL's reputation at Goddard was for high cost as well. If Goddard, a NASA-owned and -operated facility, unlike JPL, built the DMR, its civil service salaries would be carried by a separate NASA budget, thereby reducing apparent costs. If we finally settled on building the DMR instrument at Goddard or if it were contracted out, it was understood by everybody that George would become its principal investigator, the official name for the head of an individual experiment. Since it was not decided yet where any of the instruments or the spacecraft that would carry them would be made, the issue of who would be the third PI was left unresolved for the time being.

Ray Weiss came down from Cambridge to help Mike and me for the final month of writing and editing our report to NASA. He slept on the hide-a-bed couch in my little apartment at Springhill Lake, playing Mozart duets with me on my new piano for amusement and going back to Cambridge on weekends. Although he knew what he wanted to say and how to say it, English was Ray's third language. And, since completing my dissertation, I had avoided writing anything much longer than a memo.

Our difficulties were compounded by the fact that there was only a single word-processing computer available to us at Goddard. None of us scientists were allowed to work on it directly, but had to hand over our manuscripts to a single overworked secretary. We came back to her with so many corrections and inserts that she finally balked; she already was working overtime and could not handle any more. All of us lost the 1976 Christmas season in order to complete the report on time.

CRITICAL MASS

Our final product, submitted to NASA headquarters on February 1, 1977, the due date, was, unlike our slim proposal, a substantial document of four hundred pages.[8] It included several thick appendices filled with technical details and drawings undertaken for us by Ball Brothers in Boulder, Colorado. In the hope that their company's chances of winning a NASA contract would be enhanced, an engineering team led by George Yurka and Dick Herring at Ball Brothers had worked as hard as we did for a small amount of seed money. Ball occupied a compound of several buildings on the east side of Boulder, with a spectacular view of gigantic rock slabs, the Flatirons, rising up in front of the Rocky Mountains. I visited there several times and was impressed with the Ball group, all thoughtful and creative people.

One, Murk Bottema, an optical physicist from the Netherlands, came up with a novel design for the FIRAS spectrometer, although in the end we chose another. As a new civil servant, I unhappily started learning on these trips about government regulations: We were instructed not to accept any gifts whatsoever—even a ride or a sandwich—from contractors who might stand to gain from favors handed out to federal employees. I was never entirely comfortable with some of the more trivial aspects of the rules, and I confess that on more than one occasion I ate a sandwich and took a ride. Ball told me they believed they could build the entire instrument package—the three surviving instruments and a cryostat—for about $10 million. We cautiously added 50 percent to their estimate for our report.

The overall budget we contemplated came in three versions, depending on the way our spacecraft would be launched: A West Coast shuttle launch, from a facility that had yet to be built at Vandenberg, would be the cheapest, at $31.5 million, not counting the launch costs. A West Coast Delta launch would be $41.2 million, including $15 million for the expendable rocket. Most expensive would be an East Coast shuttle launch at $45 million because an extra booster stage would be required to put COBE into polar orbit. We had two ways of estimating these costs. We could ask everybody involved how much their individual pieces would cost, add in a safety factor for the things they had forgotten to include, then total it up. The second method was easier: Estimate the final weight, determine the launch vehicle category, then plot the weight on a curve with previous launches in the same category. This was brutally simple but apparently every bit as accurate as the detailed accounting method, and had the advantage of historical accuracy in hindsight.

Mike, Ray, and I went down to NASA headquarters to make our case before the peer review committee. Sitting on it were scientists from widely disparate disciplines—geophysics, atmospheric sciences, solar studies. Some had little understanding of the cosmic background radiation, why it was important, or how our proposed spacecraft and its three sets of detectors differed from the IRAS satellite. Hoping to educate them, Ray had written: "The uniqueness of the universe, the framework it sets for the natural physical laws, and our curiosity about the origin, scale and evolution of the world make cosmological studies a fundamental branch of natural science and philosophy. Cosmology, at present, is heavily weighted toward theoretical speculation because there is so little observational evidence. As a consequence, the few cosmological observables open to our investigation are of profound importance. One of these observables, accessible to direct observation only in relatively recent times, is the diffuse infrared and microwave background radiation. The COBE mission is being planned to explore the dominant contributors to that background."[9]

Ray explained to the committee how the three complementary instruments would in a sense take pictures of the universe when it was in its infancy. These baby snapshots, taken at different periods in the early universe, would give us a much clearer picture of the way the universe had emerged from this infancy. We believed we had a good vision of the scientific results we would achieve and had developed the most reasonable and cost-effective technical approach to get them. While the IRAS satellite would give us an excellent survey of current conditions in the universe, we would be looking into the deep past in the search for the very origin of the universe, our solar system, the earth, and us.

Something finally clicked: our report and oral presentation, Nancy Boggess's continuing behind-the-scenes efforts, or Ray's presentation to the committee. COBE suddenly achieved a critical mass of support and enthusiasm at NASA headquarters and, following the peer review sessions, moved to the head of the queue, just behind IRAS. Although we knew that many years of hard work and yet-to-be-defined difficulties lay ahead of us, Ray, Mike, the other science team members, and I were elated as we began contemplating actually building our satellite and launching it into orbit. Many members of the astrophysics community were less than delighted, however; in particular, X-ray astronomers who had believed their satellite would be next but whose project was now behind COBE in the race into space.

In typical NASA fashion, we learned officially of our success four months later in a brief, dry letter from Noel Hinners at headquarters: The COBE project would continue, but we were advised to keep costs below $30 million, not including the launch vehicle or data analysis. This did not seem like a problem, since all our cost estimates had been below that figure.

ORGANIC MANAGEMENT

Life was about to become very intense and exciting for a long time to come. Science team members began visiting companies around the country to see how they might contribute and whether they were interested. Ball Brothers had done a fine job helping with our proposal, but we did not want them to be our only source. George Smoot continued talking with Ball, while I opened and reopened contact with companies in the Boston area specializing in spectrometers such as Block Engineering, Idealab, Carson Alexiou, Bartlett Systems, and Honeywell. I visited the Air Force Cambridge Laboratory again, renewing my acquaintance with Tom Murdock, and also visited the Utah State University laboratory near Boston. I visited aerospace companies in California, and Mike Hauser and I attended an annual meeting of the Society of Photo-Optical Instrumentation Engineers in San Diego. Vast amounts of military funds were supporting the development of exquisitely sensitive equipment. I was not given classified details, however, and could only guess at the true state of the technology.

As I visited potential contractors, I learned the many rules and regulations that had to be sorted out. For one thing, we had to be careful about letting organizations such as JPL, where Sam Gulkis worked, or the Center for Space Research at MIT, bid competitively against private companies. Commercial contractors were sensitive, feeling disadvantaged in that universities already received large amounts of government dollars, while JPL was a FFRDC—a federally funded research and development center—and, they feared, might be given special favors. On the other hand, MIT and JPL did not want to waste time submitting bids they had no chance of winning. I never mastered—nor wanted to—many of the technicalities of government procurement.

Troubles began developing back at Goddard. This was the period when President Jimmy Carter was urging government employees to turn down the heat and turn off the lights while attempting to cut back the federal deficit. Staff members were leaving, a hiring freeze had been implemented, and in 1978 a RIF—reduction in force—was introduced. Morale at Goddard was low, with some of the engineers and staff people who had started working with us uncertain about when and whether COBE would be built, since NASA had not issued a hard start order. Despite the problems, the COBE science team was being pushed from every direction to agree to let Goddard build our three instruments and, possibly, the spacecraft itself. One official from headquarters, Leon Dondey, told us the project would be given to another NASA center if we would not agree to do it at Goddard, rather than offer it for contract. Doing it in-house would save money from the astrophysics budget.

Other cost-cutting measures also surfaced: one idea to share the purchase cost of the spacecraft with other missions, a plan that never worked out; and another of John Boeckel, head of Goddard's engineering directorate, to have the spacecraft integration work done in-house instead of contracting it out as we were planning. This was a labor-intensive process that involved making sure the instruments work together on the satellite and do not interfere with one another. Boeckel wanted to keep an engineering staff at Goddard that was at the top of its profession. To do that he constantly had to recruit young, new talent. The only way to do this was to have an exciting and challenging project for them to work on. COBE seemed to fit the bill. But how could young engineers or scientists be recruited if Goddard could not hire anyone under a freeze initiated by the administrations of first Jimmy Carter and then Ronald Reagan?

"Where are we growing our in-house talent?" Boeckel asked NASA bureaucrats repeatedly as he pushed for more civilized hiring policies that were so crucial to Goddard and to a big project like COBE.

Goddard was suffering internal pains, too, from an organizational technique called matrix management in which engineers with similar experience would be placed together in a "branch," so that for instance all the thermal engineers worked for a single manager. The concept was popular with the managers of the matrix because it gave them virtually total control over their staffs. For a project like the COBE, which still was not authorized to begin construction, it was disastrous. Under matrix management we had to approach individual branch managers throughout the Goddard complex and ask each one separately if we could borrow staff members for this or that piece of work. It meant we had to convince each manager of the project's worth in order for her or him to allow a staff member time to work with us; for us, it was a terrible diffusion of responsibility. Later on we had a cartoon depicting two canoes: a matrix management canoe with everybody paddling like mad in different directions, and a "project management" canoe with every paddler pulling forward at the same time.

Despite what seemed like endless obstacles, we managed to start pulling together a fine technical team. Goddard's engineering management discovered Marty Donohoe, who was working on another Goddard project, and asked him to become our instrument manager; they also talked Don Crosby, who had worked on infrared spectrometers for planetary expeditions, into joining us as DIRBE's instrument manager (though that was not consummated for a long time). Goddard also assigned us Jerry Longanecker and Jack Peddicord, the project manager and deputy project manager for resources, respectively, from the International Ultraviolet Explorer, which was scheduled for launch shortly.* They found Bob Maichle to work with me as the instrument engineer for the Far Infrared Absolute Spectrophotometer. Soon afterward Mike Roberto joined us as the systems engineer for FIRAS, a crucial position to fill. Mike was a technical expert whose new job would be to assure that the instrument was being designed with a coherent set of requirements and specifications. Eventually we would need a systems engineer for each of the instruments as well as for the spacecraft.

Our science team was also growing. Bob Silverberg, who had been part of our original group, rejoined us officially as a co-investigator to work with Mike Hauser on the diffuse infrared instrument. Harvey Moseley, an infrared astronomer from the Yerkes Observatory in Wisconsin who had recently received his Ph.D. from the University of Chicago, was at Goddard as a postdoc and started working with Mike on COBE. Tom Kelsall, a Goddard theorist with a Ph.D. in astronomy, also became an official co-investigator and moved into an office with me. We had sought the advice of Tom Murdock at the Air Force laboratory so frequently that we decided he should join us as a co-investigator.

Ray Weiss had met a young instructor at MIT named Edward L. Wright. Ned had graduated from Harvard and written his thesis on infrared astronomy from balloons. He was incredibly quick with computers and in the early days had actually built computers. Ray felt he would be a necessary asset and invited, cajoled, persuaded, and begged Ned to join us. He finally agreed, and Ray's foresight was rewarded as Ned became our reigning theorist and immediately proved himself invaluable in helping us avoid serious mistakes in designing FIRAS and the other instruments.

---

*After launch the IUE assumed a geosynchronous orbit over the Atlantic Ocean. Astronomers could visit Goddard or another facility in Spain and operate the orbiting telescope as easily as if they were looking through an eyepiece in a ground-based observatory. In terms of the scientific papers it produced each year, the IUE was the most productive astronomical observatory in the world.

"I was hired on as a jack-of-all-trades, and that's what I became. Everybody else had a specific job to do. But not me. I worked a little bit on everything," Ned recalled.[10] Eventually he began working almost exclusively with computers, and his remarkable ability to design complicated software programs was to prove essential.

Our growing COBE team, just like a city expanding with new population, could experience unforeseen problems, NASA feared. We were advised to create a document similar to a city charter laying out

COBE ENGINEERING AND MANAGEMENT TEAM, 1988.

Left to right, front row: Thomas J. Greenwell, Integration and Test Manager; Ernest C. Doutrich, Flight Assurance Manager; Robert T. Schools, Mission Operations Manager; Michael Roberto, FIRAS Systems Engineer; Earle W. Young, Instruments Manager; Charles Katz, Systems Engineer, Instruments; Bernard J. Klein, Instrument Engineer, DMR; John L. Wolfgang, Jr., Software Systems Manager. *Middle row:* Donald F. Crosby, DIRBE Instrument Engineer; Roger A. Mattson, Project Manager; Irene K. Ferber, Secretary; Maureen J. Menton, Secretary; Pierce L. Smith, Ground Data Processing Systems Manager; David Gilman, Program Manager, NASA HQ; Stephen Servin-Leete, Systems Engineer, DMR; Anthony D. Fragomeni, Observatory Manager. *Back row:* William D. Hoggard, Delta Liaison/Launch Operations Manager; Herbert J. Mittleman, Resources Officer; Joseph F. Turtil, Systems Engineer; Robert G. Sanford, Mission Operations Manager; Dennis K. McCarthy, Deputy Project Manager; Robert J. Maichle, FIRAS Instrument Engineer; Loren R. Linstrom, Systems Engineer, DIRBE; Jack W. Peddicord, Deputy Project Manager (resources).

areas of individual responsibility and rights for the evolving science group. Members of the original gang of six were worried about how to handle issues of scientific credit and publication, particularly since it had become apparent that many other individuals would be doing much of the day-to-day work. We also wanted new people joining our group to understand how we governed ourselves and who would be responsible for what aspects of the project.

With the help of Goddard lawyers, the team drafted a formal team charter and sent it to headquarters in December 1978 for approval by NASA lawyers. We believed we had written a fair and comprehensive document that resembled a small constitution. Unfortunately, because of the nature of the COBE enterprise, our constitution only specified appropriate behavior and did not include sanctions for violations of the letter or spirit of the charter, a failing that eventually would come back to haunt us.

In the meantime, we were still defending ourselves scientifically. Bureaucrats at headquarters wanted to know why the DIRBE experiment could not be done just as effectively from the ground. And why couldn't the DMR operate just as well and less expensively on a balloon? We were also asked why we could not run the COBE from the space shuttle like other scientific projects. Despite numerous reports and memos on our part to the contrary, some people at headquarters persisted in the belief that we should simply send our experiment up in the bay of the shuttle, point it out into space, then take down the necessary data. After all, the shuttle would be going up every week or two and would provide the simplest and cheapest route into orbit. In the excitement of the shuttle's impending first launch, we were told repeatedly that a Delta rocket, our first choice, was out of the question.*

We resigned ourselves to a West Coast shuttle launch, even though it would not occur until at least 1984 and probably not before 1985 or 1986. Headquarters also turned up the pressure on the science team to

---

*Jeff Rosendhal, our advanced programs manager at headquarters in 1979, fought for a Delta rocket, which the COBE team thought at the time was the proper launch vehicle. Had COBE been given a Delta then, however, we would not have obtained later the more sensitive DMR receivers that could be placed on a shuttle-launched spacecraft. Had we not used those receivers, the COBE team would have missed out on one of the most momentous discoveries in the history of cosmology.

agree to let Goddard build all the instruments. Building them at Goddard would give the science team easy access to the working engineers. Mike Hauser or I could simply pick up the phone or walk down the street to solve problems. Obtaining such access would be more difficult with an outside contractor because of the many legal stipulations and the big money at stake. Moreover, it was clear that we could not write a proper specification for an outside contractor. These instruments were just too far beyond the state of the art for us to be able to buy them at a reasonable cost. In either case, control of the project would be an issue, but at least at Goddard disputes would be among colleagues and friends. And NASA would save lots of money.

Sam Gulkis still held out hope that the Jet Propulsion Laboratory could build the DMR instrument, but it had become apparent that headquarters would not approve such a plan because JPL lacked civil service manpower to defray costs. We discussed the prospects endlessly. At last the science team told headquarters that we would agree to let the instruments be built at Goddard. With JPL out of the picture, the science team in October 1978 proposed the COBE's final principal investigator: George Smoot, who had wanted the job from the beginning, would now begin heading the Differential Microwave Radiometer team. George would be the only non-Goddard, non-NASA PI. Headquarters acknowledged George as PI at once, but down the road the situation would present difficulties that we had not anticipated.

One day Nancy Boggess called me from headquarters. "IRAS is in serious danger of being canceled," she told me. "It's running out of control." Its costs were approaching $100 million and its technical problems were proving increasingly difficult, Nancy said. Dick Herring, head of the IRAS work at Ball Brothers, was deeply concerned that the cryostat his company was building for IRAS would not work in space. It might vibrate too much during launch for the sensitive infrared instruments on board. If true, this would be equally disastrous for COBE, since we were dependent on the same cryostat technology Ball was developing for IRAS.

"IRAS is proving to be much more complicated than anyone thought," Nancy said. As the first large infrared astronomical space project, IRAS was encountering one unexpected problem after another. For one thing, work on the infrared detectors was shrouded in secrecy since the main contractor, Rockwell International, was also a major defense contractor for infrared technology. "We simply can't get the information we need to solve the problems as they come up," Nancy said.

The news was devastating. COBE was hostage to IRAS. We would never get a "hard start" order from NASA, much less be able to start thinking about a launch, until the IRAS mission was all but fired into orbit. Making matters worse, Nancy warned me that competing scientists from the X-ray community were circling the COBE wagon, hoping to take advantage of the IRAS problems in order to convince headquarters to schedule their mission ahead of ours, which was looking as though it would be equally complicated and expensive.

"You'd better get ready to begin moving in a political mode," she said, meaning that I should be prepared to drive downtown and make the case for COBE all over again. This time it would be behind closed doors.

As I digested the gloomy prospects, Nancy said she would see what she could do at headquarters. I felt bolstered slightly. Nancy's persuasive powers were formidable. In the ensuing weeks she went in to see James Beggs, NASA's administrator, to plead for IRAS. Fond of quoting Shakespeare, Beggs was not a scientist, but was sympathetic to scientific enterprise under the NASA flag. Beggs agreed to give IRAS more money to solve its problems.

"The ink was not dry on the check before we were back asking for more," recalled Nancy.[11] "The hemorrhaging of money for IRAS was hurting COBE and the other Explorer projects." It was suggested at headquarters that COBE was also becoming too big and expensive; perhaps the mission should be budgeted out of the Explorer program. This would mean COBE would have to seek its own separate appropriation from Congress, with all the risks that entailed. Nancy was adamant: COBE must remain an Explorer project.

Frank Martin, head of NASA's Astrophysics Division, was a staunch supporter of science who was not afraid to stand up to Beggs or to use brinksmanship tactics with him. One time, when asked to decrease the Explorer program budget, he told the administrator that, yes, it could be done and so many millions of dollars would be saved. But NASA had a unique opportunity to advance humanity's knowledge about the beginning of the universe and that chance would, of course, be lost. Martin stood his ground.

"COBE is just what the Explorer program is all about," he told Beggs and other top NASA officials.[12] In October 1981 Beggs called Nancy into his office and handed her a letter. It stated that the COBE team should start building its flight hardware the following year. The mission would be manifested for a West Coast shuttle launch on a date yet to be determined.

Nancy's magic had worked again, not only to restore the troubled IRAS project but to save COBE a second time from what appeared to be certain institutional death.

✦ 13 ✦

The Directorate

Everything should be made as simple as possible, but not more so.

—ALBERT EINSTEIN

O N NOVEMBER 22, 1980, Jane Hauser (no relation to Mike Hauser) and I were married at the Unitarian Universalist Church in Silver Spring, Maryland. We had met in New York while I was working there with Pat Thaddeus. Jane was a ballet teacher who had gone back to college to complete her undergraduate degree and get her masters degree, and had a rare combination of physical and artistic talents that intrigued me. Taking elementary math and computer programming, Jane was interested in science and had a disarming sort of intellectual curiosity; she sought to know a person's every thought, even if she did not necessarily approve of what she might find. Most of the wedding guests were involved with COBE or were Jane's ballet students.

The program, which we wrote, included our thoughts about the way each of us is constructed of atoms that originated in explosions of stars and how these atoms are recycled over and over during the life of the universe. Struggling with a word processor for the first time, I barely finished it in time to get it to the printer. The event, coupled with an argument with my future mother-in-law the night before, so frazzled me that I forgot the rings and had to ask Mike Hauser, my best man, to drive me home to retrieve them while the guests waited at the church. A friend of ours named Jack McCririe, a lapsed Baptist minister, conducted the ceremony. Ceremonial wine was passed around, and Jack blessed it.

"With this wine we acknowledge the complete interconnectedness of all things and all beings," Jack said. "We can see that these molecules of wine have been through billions of years of change, but we often forget that human beings also have a history of growth and change." Afterward everybody settled down to a potluck dinner. Jane and I socialized so much that we barely ate. Our honeymoon consisted of a one-day excursion to Lancaster, Pennsylvania.

Back at Goddard, we were attempting a marriage of another kind: combining three diverse sets of detectors and a large insulated dewar that would carry 660 liters of supercooled liquid helium into a single unit that would fly on a yet-to-be-designed spacecraft. If we could pull it off, we would have an orbiting observatory that could make measurements of the universe's diffuse background radiation wavelengths from 1 micron to 10 millimeters, a band that included the 2.7° Kelvin cosmic microwave background radiation. Two of our instruments would measure the most fundamental qualities of the cosmic background radiation, its spectrum and the way it might vary in intensity across the sky—its angular anisotropy. Definitive measurements of these, we believed, would reveal with great clarity the universe's history all the way back to its decoupling phase.

Even with the hard start status for COBE officially in the works, our troubles were far from over. We had to defend ourselves before NASA's scientific advisory committees over and over. One, the Space Science Advisory Committee, believed in COBE, while two others did not, the Space Science Board and a subcommittee on Space Sciences and Astronomy. We explained the COBE mission in a "white paper," and were allowed to continue. Meanwhile, NASA was having its own problems. The Reagan administration cut its budget twice for fiscal year 1982, and by November 1981 Congress still had not authorized funds for NASA, which operated on continuing resolutions. In the meantime, the agency's administrator, James Beggs, who was a friend of science in general and, thanks to Nancy Boggess, to COBE in particular, was under fire. His political enemies managed to have him indicted on charges relating to alleged mismanagement in his previous job with an aerospace company, and he resigned.*

The COBE project was just as unstable. Our Project Manager, Jerry Longanecker, left to work elsewhere at Goddard. Since the Project Manager would be in charge of the actual construction of the COBE spacecraft and its three instruments, this was potentially devastating. Longanecker was replaced by Roger Mattson. His deputy, Steve Paddack, departed, too, to head the Goddard office that studies future missions, and he was followed by Bob Weaver. In the midst of the departures and arrivals, we had to prepare for an encounter with the Blue Team, an independent sixteen-member committee appointed by Goddard to monitor our progress and technical designs. Rumors were floating around that the team was considering how to save money by eliminating either the DMR or DIRBE instrument.

The review lasted for days, with page upon page of official comments and questions passed around. Blue Team members told us that we did not appear to be prepared to start building COBE, that there was a shortage of engineering management people at Goddard, and that the VAX computers we were using would be obsolete by launch time. Were we going to construct our instruments conservatively? They would be strong and heavy but would overtax the cryostat, which would not keep them cold as long, possibly leading to unsatisfactory scientific results.

"Do you have any idea what kind of vibration to expect during a Shuttle launch?" Blue Team chairman Joe Fuller asked. "How will your instruments be affected?"

A mathematical model we had constructed to simulate a Shuttle launch was probably invalid, since confirming data did not yet exist, he observed. We tried for months to work out the answers. At last I wrote a confidential memo to the science team proposing thirty-two simplifications and cost-cutting measures to be divided evenly among the three instruments.

Headquarters agreed with the changes and told us to keep moving.

---

* The charges against Beggs, whom my colleagues and I considered an excellent administrator, were subsequently dropped for lack of evidence. Some observers speculated that these charges had resulted from Beggs's steadfast opposition to NASA's undertaking military projects.

SCIENTISTS VS. ENGINEERS

The science team discovered it was not easy to work with engineers. Our organizations required over eighty revisions to the formal memorandum of understanding, so it was clear we did not have a meeting of the minds. Scientists felt the designs often were full of runaway engineering conservatism, with many instances where both scientific and engineering requirements were overestimated. In other cases, scientific or engineering solutions to a problem differed significantly. To add to the confusion, Mike Hauser was spending much of his time in California working on his co-project, IRAS. The project manager's office only reluctantly accepted the authority of his deputy, Tom Kelsall. In fact, the authority of the entire science team was minimal as far as the engineers were concerned. We produced documented requirements and formal agreements, but even these often were insufficient. "The bottleneck is at the top," went a saying that applied to the science team and, especially, to me as we discovered we would be asked to define engineering requirements as well as devise conceptual solutions. The engineers were equally dismayed.

"The scientists are naive and arrogant," Anthony Fragomeni stated repeatedly.[1] "They can get tangled up in their own underwear faster than anybody."

Tony, our new observatory manager, was in charge of designing and building the COBE spacecraft. He was from upstate New York and had received his degree in aeronautical engineering from St. Louis University in 1953. Tony had worked for Martin-Marietta building aircraft, missiles, launch vehicles and an unmanned, maneuverable, and recoverable spacecraft. In 1956 at Cape Canaveral he witnessed the launch of the first Vanguard rocket; it lifted a foot or two off the ground and then collapsed into its blast hole. His domain was the entire spacecraft hardware, except for the three instruments and the cryostat. Tony had a marvelous sense of humor, which he displayed frequently with colorful use of the English language. His remarks usually were aimed at getting something done quickly, and generally worked. But just in case, he kept an "attitude adjuster" handy: a large yellow plastic baseball bat that he often grabbed to make a point.

"Should I speak directly to your branch manager?" Tony would ask the recalcitrant party. Work went faster after he arrived.

But personnel problems arose. George Smoot complained to me that the electronics team Goddard had assembled for him lacked the requisite skills. Compounding the problems was the fact that nobody knew George's exact role, since he was the only principal investigator not employed at Goddard. The PI job extended into every technical area, but as a federal agency, NASA could not abdicate control of its resources to an outside person. A deal was struck: George would "donate" his considerable expertise in instrument building to the engineering team. Dick Weber, the systems engineer on the DMR, was to draw up instrument specifications that George would review and approve. George would subsequently review the Goddard proposal written in response to the specifications. He also would contribute to instrument design and the data analysis plan and would test the instrument "breadboard," a quick and dirty lab model of the flight equipment.

Despite the procedural agreement, work on the DMR still did not go smoothly. George learned that several new kinds of microwave receivers would become available in two or three years. "With these receivers we could do the entire DMR experiment in a single day," George told the science team. All the new instruments—cooled mixer receivers, cooled superconducting Josephson junction receivers, and a new kind of transistor receivers—would require cooling. The engineers balked and stood their ground. George was deeply upset, complaining frequently to Ray Weiss about the conservative mind-set of Goddard engineers. Phil Lubin, who had worked as a postdoc with George at Berkeley and was now a COBE co-investigator, believed the new DMR instruments could work just as well from a balloon. Sam Gulkis disagreed, saying the original COBE technology should simply be updated as the launch date continued to slip. Things became so heated that at one point George threatened to drop out of the project. Ray took him aside for a long walk and convinced him to stay.[2]

Months of meetings and teleconferences were required to sort out the technical and interpersonal problems. Ray and I acted as peacekeepers, convincing the engineers to keep working with George as they looked into the new designs, while Phil Lubin tested cooled receivers. George finally agreed to an idea of Dave Wilkinson's: We would drop the lowest frequency DMR—23 GHz—and cool the 53 GHz and 90 GHz receivers to about half of room temperature. This would not require cryogenics or a refrigerator, since the receivers simply would cool down by themselves once in space. The major problem was testing them, since Goddard did not have a suitable cold test chamber yet. In partial compensation for the lost wavelength, Nancy Boggess arranged for funding to support balloon work on the 23 GHz region, a job that Dave Wilkinson and his Princeton group of Dave Cottingham, Ed Cheng, and Steve Boughn would undertake. In several flights over a number of years, Dave's group got excellent measurements at 19 GHz, but mainly confirmed that the data at the wavelength abandoned by COBE were too contaminated by galactic emissions to be useful for cosmology.

THE SPACE RACE, PART II

During the period the COBE team was designing and beginning to build its instruments, Ned Wright attended an International Astronomical Union meeting in Greece, and brought back word that the Soviets were building a DMR type of instrument. My colleagues and I were dismayed to learn that the instrument was scheduled for launch in 1983. As we studied details of the Soviet plans, we were comforted slightly: We believed our design was superior and felt the 8-millimeter wavelength the Soviet scientists contemplated studying was too long. On the other hand, if the cosmic signals did turn out to be large enough for the Soviet equipment to detect, the race would be over and we would have finished in second place.

The mission, called the *Relikt,* eventually was flown successfully. When the data were analyzed, curious sky features appeared that one could interpret as the first faint images of anisotropy in the cosmic background radiation. Looked at more carefully, the data revealed their secret: The instrument's antennas had been overly sensitive to the Moon. After eliminating the suspicious lunar anisotropies, the Soviet scientists still claimed their sky maps contained the earliest hints of cosmic anisotropy. Their map differed significantly from the ones we obtained later, and was never confirmed.* The Soviet scare demon-

---

*Years later three members of the Soviet *Relikt* team—Igor Strukov, Gena Sholomitski, and Rashid Sunyaev, one of the world's foremost theoretical physicists—visited Goddard. They were treated like visiting royalty, with tours and photo opportunities in the laboratories and a series of parties that included top NASA management and a square dance.

strated to us how nearly impossible it would be to make the measurements we were planning for COBE.

We received new word, meanwhile, that the IRAS project was in danger of cancellation because of technical failures, cost overruns, and schedule slips. We were deeply worried: IRAS was developing the cryostat technology we would use, it was developing infrared detectors we needed, and it was in front of us in the queue. Worse, testing of the IRAS dewar was not going well. If it could not maintain a low temperature, COBE's might not either. At last it was discovered that glue had come loose on one of the IRAS vent lines, and, at great expense, the cryostat was opened up and repaired.

Even the finest engineers found cryogenically cooled equipment difficult to work with, primarily because the properties of ordinary materials change as they get cold: Plastics become hard and brittle, certain stainless steel alloys disintegrate as their crystal structures change, glue pops off as it loses its adhesive abilities. You could not simply take a screwdriver and open up a dewar containing cold helium, look inside, and fix a broken wire. The dewar had to be warmed up slowly over a period as long as a month to avoid stress on the many individual parts, all made of different materials expanding at varying rates.

Most, but not all, of IRAS's technical problems were solved by late 1982, and it was readied for a dusk launch on January 25, 1983. Nancy Boggess and Mike Hauser flew out to Vandenberg to watch. Enough problems still persisted on the spacecraft, which had been built in the Netherlands, that Hans Mark, NASA's deputy administrator, believed it would fall apart under the stress of launch. He threatened to cancel it up until the very last minute.

It rained and the wind blew so hard on the afternoon of the launch that Nancy thought perhaps IRAS was jinxed. "I wondered if we would ever get it up," she recalled. In fact, IRAS came very close to not getting off the ground. Nancy was in the control room at Vandenberg when Frank Martin, head of NASA's Astrophysics Division and a strong supporter of IRAS as well as COBE, gave the final okay. "Given everything that had happened, Frank literally put his job on the line," Nancy said. "I still have the greatest admiration for his courage."[3]

Martin's gamble paid off as IRAS riding a Delta lifted off beautifully, achieving a nearly perfect polar orbit at an altitude of 559 miles. After a few initial technical problems were worked out, the IRAS observatory began delivering data as promised. The COBE team breathed a collective sigh of relief. Not only was IRAS no longer in the queue to get into space, but it had begun to demonstrate that the cryogenic technology was workable after all.

COBE slogged along. Goddard had designated it a "best effort" project. This meant that any other project was seen as more important, and engineers could be pulled off the COBE and reassigned. Although it was not in the Explorer lineup, a repair job to be undertaken by shuttle astronauts on the Solar Maximum satellite, crippled in orbit by a blown fuse, and the Hubble Space Telescope being readied for launch were ahead of COBE in scientific importance as far as NASA higher-ups were concerned. The COBE budget was flexible and so was the schedule, and I learned that COBE was at the bottom of the engineering directorate's priority list. The three principal investigators and the three instrument engineers were tearing their hair to get people assigned to the project.

"COBE was considered a good training tool by Goddard management," said Dennis McCarthy, who was named deputy project manager in 1983.[4] "People could be pulled off at a moment's notice to go work on something else." Dennis was a 1962 graduate of the University of Pittsburgh in aeronautical engineering, and had worked on a number of scientific projects at Goddard since the early 1960s. He would become one of COBE's greatest assets. At the time Dennis signed on, our launch was scheduled for 1989. But McCarthy and the other engineers held out little hope that such a schedule could be kept.

"The instruments were the problem all along," Dennis recalled. "They were all new technology, the most sensitive measuring devices ever sent into space. They had never flown, nobody was sure they would work, and they were way behind schedule."

Mike Hauser's DIRBE instrument, in particular, was troubled. Design work was going slowly, especially when Mike was away to work on the IRAS. Headquarters wondered if the instrument was necessary. Mike was worried, and Nancy went to work behind the scenes downtown to keep the instrument alive. Within a short time an instrument manager, Don Crosby, was finally appointed officially after many months of waiting, and work on the instrument began progressing. The number of wavelengths that DIRBE would measure was increased, based in part on the advice of Martin Harwit and other outside astronomers. (See figure 4.)

Mike and Tom Kelsall also decided they should measure the polarization of the light at the three shortest wavelengths because interplanetary dust radiation would be polarized at those bands. A small off-axis Gregorian telescope with mirrors but no lenses was chosen to help eliminate signals outside the field of view. Mike and Tom worried that the instrument would not be able to detect a cosmic signal in the presence of planets, the Moon, or other bright objects. They also realized the instrument would have to be tilted 30 degrees away from the spin axis of the spacecraft so that it would scan half of the sky every day.

Although not threatened with immediate extinction, FIRAS had only a handful of Goddard engineers assigned to it and nobody to conduct lab tests. Ray Weiss worked in a lab at MIT designing and building a working model of the spectrometer. Ned already was analyzing data taken by it. Ray had a remarkable idea for improving infrared detectors. For long wavelength infrared signals, the only

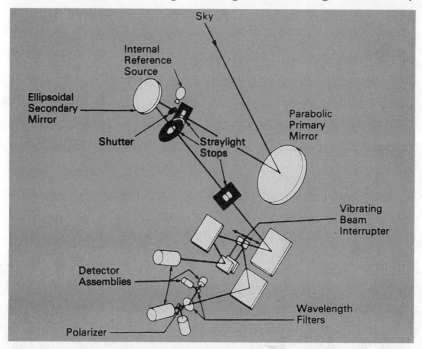

Fig. 4. DIFFUSE INFRARED BACKGROUND EXPERIMENT.
DIRBE concept, showing off-axis Gregorian reflecting telescope with 19-cm diameter primary mirror. Instrument is cooled to 1.5° Kelvin by copper straps to liquid helium tank. There are sixteen detectors of four different types.

detection method was to absorb its energy, convert it to heat, then measure the heat; this would require a thermometer of extraordinary sensitivity and an energy absorber of extremely low heat capacity in order for the minuscule amount of incoming radiation to raise its temperature quickly. Frank Low devised the old method while working for Texas Instruments. It relied on embedding tiny amounts of impurities into a chip of germanium only a fraction of a millimeter in size. The chip's electrical resistance would be dependent on its temperature because of the impurities.

Ray wanted to make the chip of etched silicon. The impurities would be blasted into it by an accelerator, with the entire structure cut to shape by chemical etching and its surface blackened with metallic film to absorb radiation. He and the MIT team, including Pat Downey, made a number of the new detectors, but found accelerator implantation of the impurities difficult to control and the final product exceedingly fragile. Worse, after all the tedious work, they found the new detectors would operate properly only at temperatures even lower than the COBE cryostat would be able to provide. At Goddard, Aristides Serlemitsos, a Greek scientist from the island of Samos (where Pythagoras lived), worked on an improved model of the original version. He used a large but very thin diamond, 7.8 millimeters wide, gluing on a piece of silicon as a thermometer with Kevlar fibers holding the apparatus in place. The detector, literally more precious than gold, worked well.

The FIRAS instrument needed bearings that would allow a mirror to move very precisely millions of times and could withstand the punishment. Roller bearings created so much friction that the liquid helium would heat up and boil away. A Goddard mechanical engineer named Meredith Wilson built magnetic bearings he believed would never wear out. They were tested extensively. Unfortunately, they were not stiff enough and their electronic controls were too complicated. Bendix manufactured metal leaf springs mounted in cylindrical sleeves called flex pivots that might do the job. Large flex pivots were used to mount the Saturn 5 rocket's engines. The ones we needed were only half an inch in diameter. Although not as strong as bearings, they held up beautifully during a rigorous pounding on a test jig an engineer named Ken Stark built. He assured us the little flex pivots would survive millions and millions of mirror strokes.

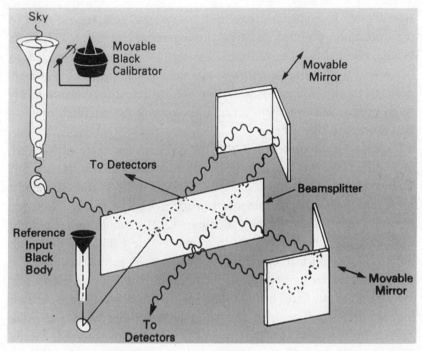

Fig. 5. FAR INFRARED ABSOLUTE SPECTROPHOTOMETER.
FIRAS concept, showing Michelson interferometer used to measure wavelengths.
Instrument is cooled to 1.5° Kelvin by copper straps to liquid helium tank. Four very
sensitive thermometers are used as infrared detectors.

We were swamped by technical challenges. One was to figure out
where the satellite was looking once in orbit. As designed, the spacecraft would have Sun and Earth sensors to control its orientation—or
attitude—to the sky. The sensors would provide orientation within
a fraction of a degree, not nearly good enough for the DIRBE
team. The on-board gyroscopes would only track the satellite's spin,
which itself would exacerbate the problem of orienting the satellite.
More than one engineer told me we could do no better with existing
technology.

Ned Wright, our project genius, came to the rescue. He had transferred to the University of California at Los Angeles when MIT failed
to offer him tenure. Considering Ned's brilliance, I was amazed. If
somebody told Ned that no solution to a technical problem existed, he

would solve it. Something easy did not interest him. His solution to the orientation problem was typically ingenious and simple: The data being collected by the DIRBE instrument itself would be used to determine the satellite's attitude. Precise measurements of the stars detected by the instrument would be used to refine the measurements taken by the Sun and Earth sensors, connecting the satellite's orientation directly to the real sky.

Even with the success of IRAS, the cryostat remained a tricky piece of engineering. We had a long list of potential failures: the support straps could break during launch, the wiring for the instruments might break, the helium might run out too soon. And what was too soon? Nobody knew the mission's lifetime. COBE would be exploring uncharted scientific territory. Its instruments were unlike any ever built before. It was unclear exactly what they would see or how long it would take to see what they might find. We thought we needed six months—half an Earth orbit around the Sun—of cryogenic operation for the three instruments to scan the whole sky. But would we really have six months? What if the cryostat failed after four months? Would we have collected sufficient data to declare the mission a scientific success? We discussed the lifetime requirement endlessly. Even though the science team wanted a year, we at last agreed on a six-month requirement.

Selecting instrument sensitivity requirements was equally tricky. The FIRAS and DIRBE instruments did not worry us. Each would have a sensitivity many times greater than any similar instrument ever built before. If they achieved only a percentage of their potential, success would be guaranteed. Settling on the sensitivity of the DMR detectors was more difficult. As the Soviet *Relikt* mission had demonstrated, the hot breath of competition was just behind us. Other groups—even the one led by our own Dave Wilkinson—were preparing similar detectors for balloon launch. The state of knowledge, moreover, for the cosmic fluctuations we hoped DMR would see was so primitive that we had little to go on.

Theorists kept moving our target. They changed the parameters every few months, it seemed, as measurements of increasing sophistication failed to find evidence of anisotropy. The fluctuations in the cosmic microwave background radiation, the theorists told us, must be even smaller than they had thought.

SPAGHETTI SOFTWARE

COBE scientists and engineers started designing the project's computer software in 1983. The problem was daunting because of the array of computers. Each instrument had its own small computer to run ground-test programs; the spacecraft would have a rudimentary computer to remember and execute commands; there would be microcomputers with a few thousand bytes of memory inside the FIRAS and DIRBE instruments to conduct preliminary analyses and reduce data before transmitting them back. Additional computers on the ground would send commands to the spacecraft, while others would receive data and place them in a specified order. Still more computers would analyze the data once they were assembled. Don Krueger, head of the Applied Engineering Division at Goddard, took the first steps by bringing together a team to design the Instrument Ground Support Equipment. This was an assortment of computers and electronic simulators into which the instruments could be plugged so they could be operated as though they were on the spacecraft in orbit.

The science team was reluctant to bring in outside help. We imagined we could solve the computer and software problems ourselves. One of our earliest solutions came with a turban: Dr. Surendra Anand, a Sikh who had been an astronomer at the University of Toronto, where he had taught several Ph.D.s, and then left Canada to find his fortune in the United States. One day he walked into my office; he told me he was seeking a starter contract for the small company he had created, Applied Research Corporation (not the large company of the same name).

"My company can locate the best astronomers to work on software for the COBE," he said confidently. "I have a large number of contacts in the field." Anand was persuasive, and succeeded in convincing me he could deliver on his promises. I already had begun an advertising campaign to find the right software person, and with contract in hand Anand began beating the bushes too. One person, Richard Isaacman, applied to both our ads. He arrived for his interview with a patch over one eye; his infant son had scratched him as he was getting on an airplane from Hawaii, where he was an astronomer on the United Kingdom Infrared Telescope on Mauna Kea. Anand and I were impressed. Rich, a Ph.D. student at Cornell who finished his degree at Leiden in the Netherlands, was smart, intense, and funny. The first person Anand hired for the COBE project, Rich began developing a team of scientist-programmers to analyze data from the FIRAS and DIRBE instruments. The DMR data also would be analyzed at Goddard, but under the direct supervision of the science team.

Don Krueger began warning us that designing the software was a greater challenge than anybody had imagined. Moreover, we had seriously underestimated its budget. Creating software was like designing a person's thought process congealed into "if this . . . , then that" statements. It was difficult to tell whether changes in a program were in terms of "scope"—that is, the goal—or just a change in how the program would get from point A to point B. The problem all along was that nobody could tell exactly what kinds of programs were needed until the actual data to be collected from the cosmos were in hand. There was little agreement even on how to construct a plan for designing our software. I favored starting on the inside of a problem at the core, working my way outward as I learned and tested my ideas with a functional, direct, organically grown system.

"This approach may have a logical flow somewhere in it," Ned said. "But it ends up as tangled as a big pile of spaghetti." Another approach, more common in the commercial world, began with defining the major requirements, then designing the overall architecture before making the interior aspects operate properly.

"We're all over the place on this software," Ray Weiss declared to a NASA committee called the Management Operations Working Group. "It's going to take us at least a year more than we thought just to get the data into scientifically usable form." The science team could not even agree on whether to use a so-called high-level command language. The scientists working on the Space Telescope had one, so a group of COBE software people paid a visit to the Space Telescope Science Institute in Baltimore. Their software would not work at all that particular day. We decided not to investigate it further.

Don Krueger also brought in outside companies to help engineer the software. One was a minority-owned business named Systems and Applied Sciences Corporation, which brought with it a big advantage: Because of affirmative action, SASC could be hired at once to get started on the software project, without going through a costly competitive process that could take up to a year. Nobody knew SASC's track record, but Don and I met with their representatives and were impressed—cautiously. They had done almost no work in astronomy, but had developed ground stations for weather satellites in the Chinese desert and complicated traffic control software for Washington. SASC also brought along for us a talented astronomer from India named Ramesh Sinha.

Ramesh was a child prodigy in mathematics who had finished college when he was fourteen years old. One day he had met a scientist on a suburban train in Bombay who suggested he become an astronomer. He earned two masters' degrees in India, and then a Ph.D. in radio astronomy from the University of Maryland. Later he wrote software for the National Radio Astronomy Observatory in Socorro, New Mexico, and had a wide network of young software specialists whom he could call on to come to work for COBE on a moment's notice. He once joked to me that this was necessary because he was not a very good scientist himself.

Convincing talented young software engineers who could earn big industry salaries to come work for a government science project with no certain future was not easy. The romance of working on software to measure the beginning of the universe helped. But once Don or Ramesh hired promising individuals, their contract could not simply state: "Go design software to measure the beginning of the universe." A new software engineer needed a formal plan and a formal requirements document in order to get started. Devising such a plan would take years of work on the part of COBE scientists and engineers.

Surprises awaited us as we began testing the new instruments. Microphonics already was a big worry. This occurred when a sensitive detector responded to vibrations of the spacecraft. The DIRBE and FIRAS detectors were suspended on tiny lead wires, and anytime a person touched the apparatus the detectors would jiggle, recording spurious information. The spacecraft would be a rigid aluminum structure that would ring like a bell if it started vibrating, either from being struck outside or from internal motion such as the rotating wheels used to control the orientation. This vibration could ruin incoming data, but would be difficult to control once the detectors were in orbit.

Another source of interference could come from the magnetic torquer bars that were part of the spacecraft's attitude control system. The iron bars were surrounded by electromagnetic coils that would switch off and on, possibly creating an electromagnetic buzz that could disrupt data. Another possible space hazard occurred to a COBE engineer.

The spacecraft would be bombarded by beams of electrons originating in the earth's geomagnetic tail. These electrons light up the polar skies in the form of the aurora borealis and the aurora australis.

"We could get hit by electrons with an energy of twenty-five thousand electron volts," Joe Turtil, who was now the COBE project systems engineer, announced one day. "The heating effect could be twenty-five watts."

This was not a small problem. If the spacecraft and instrument configuration were not designed properly, sparks could jump from one side to the other. This was a frightening possibility, since there was evidence that at least one spacecraft had been destroyed by such sparks. There was little we could do about the heating effect; data simply would not be taken during and immediately after the spacecraft passed through a magnetic storm or one of the auroras. Sparks on the sensitive instruments, though, could destroy them in an instant. Joe and the other engineers solved the problem by using electrically conductive paint and by grounding the superinsulation to the spacecraft's metal structure to eliminate internal sparking.

Another worry was the earth's magnetic field. Turning in the field, the large spacecraft would behave like an electric generator producing large currents running through it. The DMR receivers would be particularly sensitive to the earth's magnetism, since each one contained ferrite, a magnetic material that would switch the receiver's input back and forth between the two antennas. The ferrite could respond to the earth's magnetic field, creating an erroneous signal. This was solved by placing magnetic shielding around the sensitive components. As we built and tested the components of the DMR receiver and the other instruments, hoping to eliminate the producers of unwanted noise, another DMR problem developed. George Smoot complained to me that Phil Lubin was not doing his job.

"George won't tell me what he wants done," responded Phil. "We have a serious communication problem."[5] Phil asked me if he could be provided with separate funding for his work. I declined. Three of COBE's original six-member science team were DMR experts, and the project could not afford another independent investigator as far as time or money were concerned. George and Phil agreed to try to work together better. Part of the problem was that what George really needed was an assistant at Goddard to deal with day-to-day technical problems.

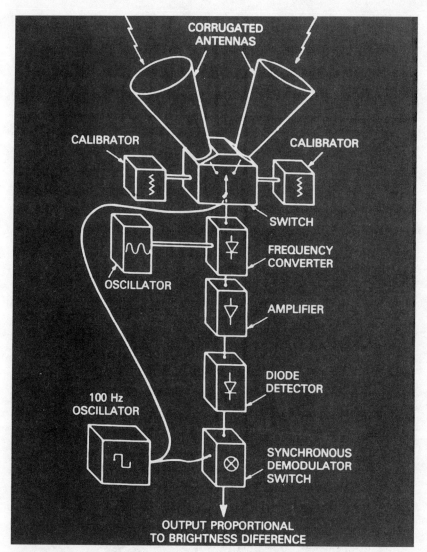

Fig. 6. DIFFERENTIAL MICROWAVE RADIOMETER.
DMR concept, showing Dicke switch connecting receiver alternately to two antennas. Output of receiver is analyzed to find the part of the signal that changes with the Dicke switch. Receivers at 31.4 GHz are room temperature, while those at 53 and 90 GHz are cooled to 140° Kelvin.

Chuck Bennett arrived in the nick of time. He had gone to MIT where he had earned a 4.0 average, then later studied under Bernie Burke, the radio astronomer who had put Arno Penzias in touch with Robert Dicke back in 1965. George interviewed him, and urged me to talk to him. I was impressed. Chuck seemed like one of those rare individuals who could work well with people while resolving technical problems. Chuck was hired into the Infrared Astrophysics section, and subsequently appointed as George's deputy PI. As a civil servant, he would be able to give official orders directly to outside contractors, an authority George lacked.

Meanwhile, the rest of the job was getting too big for the original COBE team to manage. Ray Weiss had two young postdocs working with him on the FIRAS instrument at MIT, Ed Cheng and Steve Meyer. "Ed and Steve want to base their careers on their COBE work," said Ray. "Why not make them co-investigators?" Ed had received his undergraduate degree in physics from Princeton, while Steve had been to the University of Chicago. Both had received their Ph.D.s while studying under Dave Wilkinson at Princeton. If Steve were given official COBE status, Ray said, MIT would be encouraged to grant him tenure (MIT years later abandoned cosmic background radiation as a subject, and Steve went back to Chicago). Both Ed and Steve should be invited to join, the science team agreed, but it took months for NASA headquarters to grant its official blessing. Rick Shafer, a graduate of MIT who recently had earned his Ph.D. in physics from the University of Maryland, also joined us. Rick was interested in cosmology, and his thesis involving an X-ray cosmic background sky survey made him perfect for COBE. I tried to hire him, but there was a hiring freeze on. So Ramesh Sinha signed him on to work on software. Eventually NASA hired Rick as my deputy principal investigator for FIRAS.

Every component of every instrument was being tested and retested prior to assembly. Unexpected problems arose over and over. A technical hazard in the mirror mechanism of the FIRAS appeared: It was insufficiently stiff, and could begin vibrating. Mechanical fixes did not work. The problem eventually was solved electronically by adding a "notch filter" to prevent the mechanism from sending electrical signals at a frequency that might cause oscillations. Also, the springs in the mirror mechanism that functioned as bearings, it developed, broke during launch vibration tests. A special "caging" device was constructed to hold them firm during vibrations. In California Sam Gulkis began creating a computer-generated model for the response of the DMR to the Moon in order to eliminate the same kind of problem that plagued the Soviet *Relikt*. The DMR team hoped to use the model to determine where the instruments' two antennas were pointed and to calibrate the DMR's response to an object of known intensity like the Moon.

Goddard purchased an instrument test dewar for the FIRAS and DIRBE since the actual flight dewar would be far too precious to use as a testing device and, in any event, had not arrived yet from Ball Aerospace. The test device was a blue cylinder 12 feet tall and about 7 feet in diameter. On top was a window through which Bob Maichle and Mike Roberto, the engineers leading the FIRAS team, could shine a far infrared laser signal to check out the instrument. Cryogenics testing was remarkably difficult.

"Sometimes we all just stand around hugging the cryostat, wondering what's going on inside," had said Marty Donohoe, our first instrument manager. Hundreds of wires and cables, optical fibers, electrical black boxes, and thousands of other components were tested. During cryogenic testing the FIRAS calibrator, consisting of several pounds of epoxy, broke when shaken. The design was reinforced. Still, almost everything failed its first test. By the end of 1985 the instrument test failure rate was 100 percent.[6] By then our second instrument systems manager, Earle Young, was on the job. He had his hands full, but he coaxed, prodded, urged, pushed, planned, organized, and did what was needed to pull through. More experienced than the younger engineers leading the three instrument teams, he showed them how to work with management to get the needed expertise assigned to the team. We would never have made it without him.

✳ 14 ✳

Skunkworks

Let us not unlearn what we have already learned.

—DIOGENES

THE COBE SPACECRAFT, minus its three instruments and dewar, was all but complete by January 1986. Tony Fragomeni and his engineering team had designed it to fit into the bay of the space shuttle, technically known in NASA jargon as the Space Transportation System, or STS. Squat and cylindrical, the COBE observatory consisted of two sections: the spacecraft itself and the instrument module. The spacecraft was constructed of machined aluminum panels riveted together; it was 18 feet long, 15 feet wide, and, when fully loaded with the instruments, cryostat, and hydrazine fuel, would weigh 10,594 pounds (technically, it was allowed to weigh up to 11,500 pounds). Four trunnions would fix the COBE to the bay walls of the shuttle during launch.

"It was a piece of cake," recalled Dennis McCarthy, the deputy project manager. "The spacecraft was no different than any of the others all of us had worked on. Its parameters were given. It was not pushing technology, like the instruments."

The dewar would soon be installed within the satellite's walls. The FIRAS and DIRBE instruments would be inside the dewar, with the DMR mounted on three stands surrounding the dewar. Twelve solar panels surrounding the craft were designed to gather radiation from the Sun and convert it photovoltaically into usable electrical power. A thermal shield was installed inside the solar panels to protect the dewar and DMR from the Sun's searing heat. Tony's engineers had come up with a relatively simple and gentle hydrazine thruster system to propel the COBE into an orbit some 350 miles higher than the space shuttle's upper limit of about 200 miles. There the satellite would be free from all but the most minuscule traces of the earth's polluting atmosphere.

As of the evening of January 27, 1986, almost all the essential tests of the primary structures had been done, along with 80 percent of the work on the complex electrical "harness." The harness was the nerve system of the spacecraft's complex electronics, consisting of miles of wires and cables connecting the many boxes of electronics. Before they left work for the day, Tony and Dennis formulated plans to move the COBE spacecraft, minus its instruments, into a clean room for final integration making sure all the parts worked smoothly together. The next day, a few minutes before noon, Tony sat in Dennis McCarthy's office eating a brown-bag lunch. They were discussing their plans to integrate the electrical harness with the spacecraft structure once cleanroom testing was complete. A young secretary burst into the room with a look of horror on her face.

"My God, the shuttle has exploded!" she said. At first Tony thought she was joking, but her expression convinced him she was not.

At the moment of the *Challenger* explosion, Mike Hauser was on a flight from Washington to California to attend a meeting on IRAS data analysis. He learned of the disaster when his plane arrived in Los Angeles and he found the terminal's television sets jammed with viewers. Ray Weiss was leading a committee meeting for NASA at the Pasadena Civic Center. The topic was "major new directions for the space program in the era of the space station." Ray's committee was to make recommendations on fundamental physics research to be carried out in space. Peter Bender, a member of the panel, rushed in before the morning session and told Ray that he may as well cancel the meeting, since a more important event had just taken place. George Smoot was driving from his home to Lawrence Berkeley Laboratory at the moment of the accident, learning about it upon his arrival at work. Harvey Moseley was in a cryogenics laboratory in the basement of Building 7 at Goddard conducting calibration tests on the DIRBE instrument. He had had a sinking feeling that morning, as he often did before a manned launch, even saying to a colleague, "I hope NASA doesn't blow up the schoolteacher."

Harvey was deeply aware of the complex engineering problems that had to be overcome for each successful rocket launch. Knowing the laws of probability were not operating in the shuttle's favor over the long term, he had refused to watch the launch on television. Somebody told me about the accident as I walked down a hall in Building 21.

That NASA had sent a senator and a congressman and now a schoolteacher into orbit aboard a rocket-powered spacecraft was, I believed, the height of hubris. NASA officials apparently had started believing their own public relations campaign that the shuttle was little more than a rocket airplane anybody could board for a ride into space. To the contrary, I believed it was a technological triumph each time a research project like the space shuttle operated the way it was designed to. Expecting such a complicated monster to do it right over and over— and inexpensively at that—was, I believed, insane. Like Harvey, I had not been able to bring myself to watch the launch. Throughout the rest of the day some Goddard employees watched replays of the disaster on television monitors. I did not. Nor was I yet aware of the nearly calamitous impact the *Challenger* explosion would have on COBE.

According to the public record, STS 51L, named *Challenger,* exploded after lifting off from Launchpad 39B at the Kennedy Space Center at 11:37:53 A.M. Eastern Standard Time on Tuesday, January 28, 1986. On board were seven crew members, including an African American, an Asian American, a female astronaut, and a schoolteacher named Christa McAuliffe. Highlighting the diversity of the crew and McAuliffe, NASA's prelaunch publicity campaign was so successful that millions watched the final countdown, ignition, liftoff, and then, with the rocket ascending nearly out of sight, the terrible explosion sixty-seven seconds after launch. Like the day President Kennedy was assassinated in 1963, most people probably have not forgotten where they were when the *Challenger* exploded.

Within three months a huge Titan 34D loaded with a billion-dollar military satellite exploded during launch from Vandenberg, destroying the launchpad and spewing noxious gas for miles. Dozens of people were treated. Two weeks later an unmanned Delta rocket, NASA's first launch since the *Challenger,* lifted off from the Kennedy Space Center with a multimillion-dollar GOES-G weather satellite on board. A minute into its flight the rocket's main engine lost power and, spinning dangerously, the rocket and spacecraft were destroyed by a signal from the ground.

Emerging details of the *Challenger* disaster damaged NASA's image even more: the decision, apparently motivated by public relations concerns (the real story is much more complex), to fire the shuttle's rockets on an unusually cold Florida day against the advice of several launch engineers, the leak in one of the solid fuel rockets visible later on close-up videotapes, and the eventual recovery of shuttle fragments and the crew's bodies from the Atlantic Ocean. Who can forget Richard Feynman's televised explanation of how the rocket's O-rings might have failed in the cold weather?

NASA was crippled. As depicted on television and in the print media, the public record did not fully convey the intensity of the shock waves from *Challenger* and the other accidents reverberating through the agency. With the shuttle grounded indefinitely and the entire aerospace industry in chaos, the nation's ability to launch any kind of rocket, manned or otherwise, now seemed seriously compromised. At NASA headquarters Nancy Boggess, bright-eyed guardian angel of IRAS and COBE, was devastated. Although the IRAS satellite had generated beautiful data, everything else seemed to be falling apart. James Beggs, a friend of science and one of the most approachable in a string of NASA administrators as far as she was concerned, had been indicted and forced to resign. Every project she had been working on looked as though it would be pushed back at least a year due to the loss of the shuttle. Nancy believed there was little she could do to make a difference anymore.

"The ratio of fun to frustration equaled one," she said.[1]

She began thinking of leaving NASA and started contacting influential members of the infrared astronomy community: "Please try to find somebody to take my place here," she pleaded. Knowing her frustration, Mike Hauser said to Nancy: "Why don't you come out to Goddard and work on COBE full time?" Nancy had been outside an academic environment for years and was not a publishing scientist, but I didn't hesitate for a moment. Nancy had always been a voice of wisdom and had a combination of perpetual optimism and intense focus that made things happen.

Nancy pressed the hunt for a successor to represent the infrared community at headquarters. After a lengthy search, Fred Gillett, an infrared astronomer at the Kitt Peak National Observatory near Tucson, Arizona, agreed to take the position. When she officially moved out to Goddard, Nancy took over my part of COBE's project data management. "It was a wonderful opportunity for me to get back into real science," she said. "I had been John Mather's boss, and now he would be mine."

But like everybody else on the science team, I had begun to wonder: Would there even be a COBE after *Challenger?*

DAY ONE, ALL OVER AGAIN

The COBE project was blessed with a succession of saviors at critical moments along the way: Noel Hinners, Frank Martin, James Beggs, and Nancy Boggess at NASA headquarters; Pat Thaddeus, Ray Weiss, and Mike Hauser, among the scientists. Now it was Dennis McCarthy's turn. Many scientists and engineers at Goddard and throughout NASA initially, and in retrospect, naively believed shuttle flights would continue as scheduled even from the West Coast, despite the horrifying *Challenger* explosion. If so, there would be little effect on COBE, already manifested for an STS West Coast launch in 1989. Dennis was one of the few people who recognized what the disaster would mean to COBE: "I watched it over and over on television like everybody else that day. I couldn't believe they didn't have parachutes or some other kind of escape system," he recalled. "NASA had only three shuttles left. It was not hard to figure out that there would never be a West Coast shuttle launch. I knew we had a serious problem on our hands."[2]

The day after the explosion, McCarthy arrived for work in Building 7 about 8:30. Most of the engineers assigned to COBE were at their desks. He walked down the hall, speechlessly summoning them one by one into his office. "Everybody knew something was up," said Tony Fragomeni.

"The shuttle will never launch from Vandenberg now," McCarthy said. "If we want to save COBE, it's going to be up to us to find a different way of doing it. We're paid to come up with solutions, so let's do it."

There was not a happy face at the conference table, but Dennis's manner was so purposeful that the individuals in his office began broaching new ideas and suggesting alternative approaches. "We left the meeting thinking we really might be able to do something to keep COBE alive after the awful disaster of the *Challenger*," said Tony. The first order of the day was to halt all work on the COBE shuttle-configuration spacecraft that Tony and his team of engineers had all but completed. Fortunately, they did not have to stop working on the electronic harness with its complex array of black electronic boxes. The engineers thought they could integrate the black boxes for another spacecraft if necessary. That was the good news. We did go ahead, and it saved a lot of time later.

The hard part was to find as quickly as possible another rocket to get COBE into space. Acquiring a new vehicle was a major problem. Because of its commitment to the shuttle as the one and only vehicle it would use to place satellites in orbit, NASA was in the process of phasing out its inventory of expendable rockets. Only a few remained: thirteen Atlases, six Titan 34Ds, three Atlas-Centaurs, four Titan 34D7s, and one Delta. With the shuttle grounded, competition for the expendable rockets would be intense.

Dennis moved quickly. His engineering group established two ground rules for getting COBE into space: First and foremost, science objectives could not be compromised if at all possible and practicable; second, and nearly as important, the engineering group would attempt to maintain the same launch schedule. The second point would keep costs in check and help sell a new COBE flight plan to headquarters.[3] As a corollary, the engineers would do everything humanly possible to save as much of the complex and expensive electronic system, already installed on a wooden mockup of the spacecraft, as they could.

Within a month McCarthy and Roger Mattson created a comprehensive matrix chart in his office depicting the pros and cons of various launch vehicles, domestic and foreign. Although both were powerful enough, the Titan 34D and Atlas Centaur were eliminated at once, since no launch facility existed for them at Vandenberg. The huge Titan 34D7 could carry more than 15,000 pounds into polar orbit from the West Coast with a configuration close to that of the shuttle, so COBE would have to be modified only slightly for a Titan 34D7 launch. But its cost was prohibitive: approaching $200 million for the launch alone. NASA headquarters would never agree to such an expenditure for a scientific satellite. The only remaining NASA rocket was the single Delta. It could be launched from Vandenberg relatively inexpensively, but it could carry only about half of COBE's weight. A lighter satellite would have to be built.

"The Delta was the toughest launch vehicle for COBE," said McCarthy. "It would take a complete overhaul of our spacecraft. We kept on looking."

McCarthy started thinking again about the shuttle. In April he was proven prescient when the COBE team learned through the grapevine that the Pentagon was closing down, probably for good, construction of its shuttle launch facility at Vandenberg. The military required immediate West Coast launches to attain orbits over the Soviet Union and could wait no longer for NASA to decide the future of the shuttle in the wake of *Challenger*. Pentagon officials planned to shift funds at once into a new expendable launch vehicle, an advanced and larger Titan rocket.

Technical reasons also dictated abandoning the West Coast shuttle site: It was discovered that hydrogen gas from a shuttle launch could accumulate in the valley and explode, and that the shape of the valley might reflect sound waves from the engines back toward the shuttle, exceeding its tolerance for noise and vibration. NASA also began to express fears that Vandenberg might lie too close to an earthquake fault zone for a safe manned rocket launch. Complicating matters further were Indian gravesites in the vicinity of the launch pad.

But what about an East Coast launch once shuttle flights resumed? Technically a shuttle might be fired into a polar orbit from the Kennedy Space Center, although this was not likely for practical and political reasons: A northbound launch would fly directly over the heavily populated Atlantic Coast region of the United States, while a southbound shuttle would have to be launched over Cuba. McCarthy and his boss, Project Manager Roger Mattson, had another idea for the shuttle and asked to meet with the science team. They brought along Bob Farquhar, a scientist who was an expert in satellite orbits.

"What about sending COBE out to L2?" Dennis asked us. L2 was Lagrangian Point No. 2, a spot about a million miles farther out into the solar system than Earth where the gravity of the Sun, Earth, and Moon balance perfectly.* Suspended gravitationally at L2, a satellite could

---

* These points were named after Comte Joseph Louis Lagrange (1736–1813), a French mathematician and astronomer known for his calculations of planetary orbits, who was honored by Napoleon and later buried in the Panthéon. Lagrangian points are sometimes called libration points, since an object would librate, or oscillate slowly, across them.

remain there indefinitely with occasional nudging from a small thruster. The science team "literally drooled" when they heard the idea, McCarthy recalled.[4] Understandably. Lagrangian Point No. 2 was a perfect environment for the COBE: benign, free of interfering radiation and electromagnetism, and situated perpetually in sunlight to power its solar cells. Dennis and Roger had developed an engineering scheme that seemed at least feasible. COBE, as presently configured with modifications, would be launched on a shuttle flight into a normal equatorial orbit at an altitude of about 200 miles. From there, COBE would ride a new upper-stage booster rocket built by Orbital Sciences Corporation out to L2, where it would remain.

Unfortunately, as he undertook a cost analysis Dennis discovered the new OSC rocket alone would come to $200 million. There was little point in broaching such a figure to headquarters. NASA was under fire, its budget was tight, and most of what money remained was being pumped into restoration of the shuttle. Older Goddard employees were being asked to take early retirement. They could not be replaced because of the Reagan administration's hiring freeze. It was becoming clear that we had seriously underestimated the amount of testing the three instruments would require. Our costs were rising with inflation, and as it started looking as if the COBE project would take forever or simply die of attrition one day soon, morale began to plummet.

The instruments were troubled, too: The blades on the DIRBE light-chopping mechanism were running too hot, and the FIRAS mirrors were vibrating. The software was behind schedule, and an "us versus them" mentality had emerged between the scientists and engineers in the depressing wake of *Challenger.* The COBE team's unstated but universally understood policy was to avoid confrontation with headquarters until the official strategy regarding the shuttle was announced. The best guess was that COBE would be delayed at least a year with an additional cost of $6 million to keep the project alive. Worst of all, we were back to square one. COBE was a spacecraft without a means of getting into space. The COBE team began making plans to complete assembly of the spacecraft in 1988, then place it in storage and hope for a launch sometime in the early 1990s.

Dennis went to his rocket matrix chart. He looked at two unlikely candidates: the Soviet Proton and the Chinese Longmarch rockets. Both were powerful enough, but he knew headquarters would not contemplate launching a NASA-built satellite aboard the rocket of either of our nation's two major adversaries. A better solution was the French-built Ariane. From its launch facility in French Guiana, the Ariane could easily carry into polar orbit the existing COBE spacecraft with very little modification. Its major disadvantage would be its approximately $80 million launch fee.

McCarthy thought of a way he might be able to eliminate the cost of an Ariane launch. Using a pay telephone in order to place the call as a private citizen rather than a NASA employee, he contacted the Ariane marketing office in Washington and asked whether they would like to look into an arrangement in which the launch fee would be waived in exchange for a share in the scientific data generated by COBE. After more calls and three trips with Roger Mattson to the Ariane office, McCarthy had his answer: The French were definitely interested in a deal. McCarthy was ecstatic. Of all the rockets, the Ariane would require the least reconfiguration of COBE. And now it could be free. "I wanted the Ariane," McCarthy recalled. "That was my best solution."[5] But he and Mattson were uncertain how to approach NASA with the idea.

McCarthy decided to go public, starting a rumor among the NASA bureaucracy to test the waters. "It traveled at the speed of light," Dennis recalled. Within two weeks Douglas Broome, program manager at headquarters (now deceased), called him. "This is a direct order, Dennis. Don't ever discuss the Ariane again," Broome said, his voice trailing off threateningly. Broome intimated that the command came from the highest echelons of the federal government.

Complying, Dennis retrieved the next day one last idea he had roughed out on the back of a napkin shortly after the *Challenger* disaster. The sketch depicted a COBE spacecraft reduced in size, with deployable solar panels folded around its flanks configured for launch aboard a relatively small Delta rocket. He brought it to Bill Keathley, Goddard's director of projects. Keathley told Dennis to come back in forty-eight hours and tell him whether COBE could be redesigned for a Delta. McCarthy responded: "I'll take the forty-eight hours, but the answer will be yes."

Meeting with his project office, McCarthy explained the idea. Jim Woods, the lead mechanical engineer, came up with a concept that would accommodate all of COBE's original scientific instruments and electronics. The hydrazine propulsion system would be jettisoned altogether, along with most of the aluminum structure designed to fix the spacecraft to the shuttle bay during launch, eliminating as much as 4,000 pounds. Still more weight savings would be necessary since a COBE Delta launch could weigh only half its original 10,500 pounds. Dennis said he thought NASA would give us the funds if the COBE team could show that it could quickly reconfigure the spacecraft. I held my breath.

McCarthy and Keathley drove down to headquarters, where they paid a call on Sam Keller, NASA's deputy associate administrator. "We want to get a high-profile science satellite up as quickly as possible to help restore the public's faith in NASA," Keller told them. COBE's competition was a weather satellite the National Oceanic and Atmospheric Administration wanted to launch. McCarthy showed Keller a drawing (no longer on the napkin) of COBE redesigned for a Delta.

"Can you do it in twenty-four months?" Keller asked. McCarthy believed it would take at least thirty-six months. "How about thirty-one months?" he asked. "Okay," said Keller, "you've got twenty-nine months." McCarthy left the meeting soaked with perspiration. Twenty-nine months was possible, he thought, but only if the clock did not start ticking until the beginning of the new fiscal year in October.

McCarthy and Keathley also stopped in to see Frank McDonald, NASA's chief scientist, who was proving to be a guardian angel for COBE. McDonald later stopped by Nancy Boggess's office to make sure she was on the same wavelength with McCarthy and Keathley. She was. In September Nancy Boggess learned that funds would be available for making the modifications to COBE that would enable it to fly on the last remaining Delta rocket. But she had no idea where the $40 million needed for the launch itself would come from. On October 1, 1986, headquarters sent out to Goddard firm word that launch funds had been found in the comptroller's kitty. COBE was manifested for a Delta launch at Vandenberg in twenty-nine months: February 1989.

My colleagues and I were ebullient. COBE's engineers and supporters at headquarters had found the pathway into space.

THE TIGER TEAM

The most exciting, intense, and frustrating period in COBE's history began. The dewar had arrived at Goddard shortly before the *Challenger* accident. Ball Aerospace transported the 1,426-pound cryogenics tank from Boulder, Colorado, in three specially fitted United Van Lines trucks—two hauled ground-support equipment; the third was loaded with the dewar suspended in a white, 7-ton container (which also had been used to ship the IRAS dewar), and was fitted with shock-detection equipment and an accelerometer to measure g-forces on the precious cargo during the trip. The trucks set off down Interstate 70 at precisely 54 miles per hour. Olof "Bud" Bengtson, the Ball engineer who had overseen the dewar's construction, accompanied it on its trip east.[6]

The driver of the truck bearing the dewar was apprehensive that he would not be able to give it a soft enough ride. In Limon, Colorado, Bengtson checked the accelerometer charts. "Hey, you're doing pretty good," he told the driver. "Let's keep rolling." The heaviest load on the dewar during the 1,700-mile trip was 0.9-g, roughly equivalent to a sudden doubling of its weight. That was not bad. The launch would be much rougher.

Once safely at Goddard, engineers viewed the dewar's weight and dimensions as "absolute givens" as they went about the job of reconfiguring the spacecraft. It was a cylinder with a diameter of 5 feet, 6 inches and a height of 7 feet, 3 inches and had taken years to construct at a cost of $12 million. NASA would never agree to a new one. The other absolute parameters were the launch capacity of the Delta rocket and its physical dimensions. Mattson, McCarthy, and Tony Fragomeni, who was directly in charge of the spacecraft's overhaul, tried to galvanize their troops.

"One of the problems we had at first was that some of the engineers didn't think it could be done, but we soon disabused them of that notion," said Tony, master of the attitude-adjusting bat. Getting COBE into orbit was now Goddard's No. 1 priority and one of NASA's top priorities in the absence of shuttle flights. In early 1987 NASA administrator Jim Fletcher visited Goddard and looked over the COBE hardware, then issued a press release stating that COBE was the centerpiece of the agency's recovery.

Mattson wanted to wrest control of the project from upper management in order to bypass the work-bedeviling matrix management system. He sought and received permission from Noel Hinners, now Goddard center director, to establish a "tiger team." This was NASA jargon for a crash effort in which all the designers and engineers would be situated in one building for as long as it took to complete the project. Old-timers like Tony and Dennis called it a "skunkworks," the name of a similar crash program devised by Lockheed engineers to design and build aircraft quickly during World War II.*

With work progressing, the clock began ticking faster than ever toward the early 1989 launch date. Spacecraft redesign engineers worked around the clock, seven days a week, in Building 7. Dennis McCarthy's office, at the center of the maelstrom, seemed like the site of a single, ongoing engineering meeting. To make way for the new COBE space-craft in the huge spacecraft assembly laboratory, the old shuttle-configured craft was moved to a storage room in Building 16. Its hydrazine propulsion system was sold in bits and pieces to an Air Force SDI mission and eventually to a joint Japanese-American mission to study rain forest depletion from space. Although the money should have gone into the general treasury, somehow it found its way directly to the skunkworks.

"It was a bonanza," Tony said. "We spent it directly on rebuilding COBE."

Morale soared in the face of the challenge. "The change from STS to Delta required a complete change in the grounding system," said Mark Flanegan, COBE's command and data handling systems engineer. We decided to use the dewar itself to hold up the DMR, saving about 60 pounds of weight.[7] (See figure 7.)

Every pound was crucial as the engineers struggled to cut the spacecraft's weight from 10,594 pounds to at most 5,025 pounds and its launch diameter from 15 feet to 8 feet. Loss of the propulsion system, not needed since the Delta rocket and its boosters would carry COBE all the way to the required altitude, saved 2,064 pounds. The new, leaner spacecraft structure would weigh only 1,000 pounds, a savings of 3,803 pounds. The new electrical harness would save 204 pounds, while the smaller solar panels and power system would weigh some 300 pounds less. The three instruments—FIRAS, DIRBE, and the DMR—would, of course, weigh the same, 1,142 pounds, so no savings would be obtained there.

Eureka!

---

*The exact origin of the name *skunkworks* is a mystery, but it may have come from the presumed odor resulting from many people working hard together under one roof. Mike Ryschkewitsch thinks it came from "Li'l Abner."

Fig. 7. COBE SHUTTLE CONFIGURATION.
COBE was proposed in 1974 for launch on an expendable Delta rocket, and later redesigned for launch aboard the space shuttle. Here is how COBE might have looked had it been launched by the shuttle.

After months of redesign and endless laboratory tests of new systems and subsystems, Tony's team calculated that the new spacecraft would weigh in at exactly 4,828 pounds. They had trimmed 5,766 pounds of fat, with a 197-pound margin of safety for good measure. Moreover, its dimensions had been reduced just enough to allow it to sit precisely atop a Delta rocket, primarily by replacing large and heavy built-in structural solar panels with solar cells mounted on lightweight hinged panels that would deploy in orbit.[8]

"This raised thermal questions," said Rob Chalmers, the spacecraft's thermal engineer. "As the wings swung between the Sun and the spacecraft, what impact would that have on the temperature of the components? Would the glass-covered cells concentrate sunlight onto the spacecraft?"[9] After conducting spectral analyses and struggling with archaic specular modeling software, Chalmers told Tony he thought that the effects would be minimal.

Goddard was mobilized as never before in its history. A team spirit evolved, despite the fact that many engineers worked double, and occasionally even triple, shifts. Old worries about overtime pay were discarded as Goddard management pushed to make sure we met head-quarters' launch date. Expediters appeared to make sure the schedule was kept, documentation and reporting were streamlined, and purchases were immediately processed.

Some 300 Goddard engineers, designers, scientists, and contract support staff worked on COBE at the height of the rebuilding effort in 1988 (although many still worked on other projects too). The full list of people who ever worked on COBE is given in the appendix. Counting everybody, there were about 1,600. As the pressure increased, tempers flared between exhausted engineers and lab technicians, while old animosities between engineers and scientists occasionally resurfaced. Ray Weiss wrote a stern letter to Roger Mattson, chiding him and his engineers about what Ray considered an amateurish job on a new electrical circuit.

"He told us that one of his students could have designed a better one," Tony said. "We looked it over carefully, and decided it was okay in spite of what Ray thought."[10]

The three instruments were the least of the worries in the skunkworks office. The FIRAS and DIRBE detectors would dwell inside the dewar, so no reconfiguration was needed. Only the DMR, mounted outside, would have to be changed. The original COBE was big enough to allow the radiometers to be mounted on pedestals sufficiently removed from the dewar to make interference between the instruments minimal. The Delta redesign would have to fit inside a tighter envelope without causing excessive interference between the instruments. Gene Gochar, a skunk team mechanical engineer, headed a new DMR design that would accommodate existing electronics, satisfy thermal constraints, and fit within the smaller spacecraft. The new housing became wider and shallower, while the lower part was curved to fit around the upper part of the dewar. With construction of the new DMR housing under way, testing of the original radiometers went forward. We also continued the endless tests of FIRAS and DIRBE and their most minuscule component subsystems.

The tests were not going well. Mike Hauser reported that the light choppers on the DIRBE still did not work and, worse, the manufacturer was giving up on them. Casey DeKramer took over the job and was dispatched to the corporate laboratory to collect completed parts. Black paint on the DIRBE baffles was also causing trouble, chipping in vibration tests. Any paint that chipped off in orbit would interfere with the instruments. We took a calculated gamble and left off the paint. The DMR was also troubled. Undergoing testing in a clean room, a supposedly dust-free environment in Building 19, the front of the receiver that converts incoming signals to lower frequencies in order to be amplified was not working well. We appealed to Hughes, the manufacturer, for help, even though most of the money NASA had agreed to spend was gone. Hughes officials responded nobly, making repairs at their own expense.

In the nick of time John Wolfgang, an electrical engineer who had been building spacecraft at Goddard for nearly twenty years, decided to come aboard as integration and test manager. "I was working on something else that was going pretty slowly, and thought, hmmm, this COBE looks like it's pretty interesting."[11] A Pennsylvania native, John had graduated from George Washington University and developed a range of interests in the Washington, D.C., area, teaching computers at a local college and running an Explorer post for space exploration while still managing to work many overtime hours at Goddard. Along with a round shape and gruff manner, John brought immense determination and skill to COBE. At his first meeting with the science team, he listed thirty-five items we needed to fix or change at once.

Every organizational boundary was an invitation for confusion, I had learned by now, and we often were as poor in communicating as we were rich in organizations: three instrument hardware teams, three instrument software teams, the project management staff, two major contractors, government and contractor organizations, and all the universities that were represented on the science team. Problems with interorganizational software, especially, escalated.

"The system is as disorganized as the tower of Babel," Ray Weiss declared at one of the many review sessions. For their part, the software engineers stated accurately that the scientists had failed to supply them with a complete list of requirements. After one in a series of endless meetings, one software review committee member walked out with "a very sick feeling," as much over the scientists' lack of experience as over the lack of progress in COBE's software development.

Trouble among outside software contractors also arose. ARC, Anand's company, was financially troubled and apparently owed a large sum to the Internal Revenue Service. After solving that problem, Anand immediately was beset with another: Rich Isaacman, his top software scientist, left to join a different company, General Science Corporation, just as a formal competition for a contract was under way. We needed Rich's expertise on COBE, but contracting rules prevented him from working for us in the same capacity. After long negotiations, peace of a sort was achieved: ARC and General Science agreed to team and split the outside software work.

Testing was endless, each new test seemingly more vital than the last and usually yielding poor results. The FIRAS mirror transport mechanism still balked, the external calibrator was troubled by the cold inside the dewar, and we began to fear that the DIRBE tuning-fork chopper mechanism might destroy itself during launch. Under certain circumstances, the second stage of the Delta rocket would vibrate at 30 to 35 Hz. Since the DIRBE tuning fork was designed to oscillate at 32 Hz, it could shake itself to bits under the stress of additional vibrations. After many tests and meetings, a system of bumpers was devised to prevent the tuning fork from moving much during launch. A new FIRAS mirror transport mechanism was brought in, meaning more tests were required. Without the mechanism, there would be no FIRAS, no COBE, and no measurement of the beginning of the universe. Mike Roberto, FIRAS's systems engineer, constantly performed heroic acts to ensure that the tests were productive.

With COBE's schedule slipping and its budget rising, pressure from headquarters mounted. At a large biennial review session attended by NASA's top brass, Charles Pellerin delivered what I construed as a threatening pep talk: Congress had been told that COBE, NASA's No. 1 priority, was to fly by May 1989 or else. The obvious threat was that the whole management could be fired if we failed, though I never heard it said out loud. Everybody at headquarters and on Capitol Hill was watching us closely. To my relief, Frank McDonald, NASA's chief scientist, spoke up in COBE's defense, stating that the emphasis should be on a successful mission rather than a fast one: "No one ever won a medal for an on-time failure."

John Wolfgang, overseeing the integration and test program in Room 172 in Building 7, which was filled with computers and people day and night, seemed to have an uncanny depth of knowledge about every last nuance of the engineering test program. I was simply amazed by John's knowledge of every detail. I learned only much later that he had developed a secret method for getting the information out of his crew: He kept a small refrigerator in his office. At the end of a shift, an engineer or a technician would be invited into John's office where, over a beer or two, John would pry out of the individual every last detail about the day's work. Denis Endres was John's right-hand man, making sure everything went according to plan.

Engineering staff members work on the COBE spacecraft in a clean room in Building 7 at Goddard Space Flight Center several months before launch.

An ever-present problem was eliminating interference among the instruments as they were brought closer together in the shrunken version of COBE. "What makes the job so hard is that we're often trying to find noise within the noise," said Bob Martin, one of the engineers. Tony Fragomeni likened the ongoing instrument noise problem to that of "picking fly specks out of black pepper."

The spacecraft itself resided a hundred feet away from the engineers in a clean room at Goddard. In September 1988, it was time to fill the dewar with liquid helium. Everybody gathered on the appointed day; endless discussion ensued. "I counted thirteen Ph.D.s from the science team and the cryogenics group along with a number of engineers and managers around the spacecraft trying to decide whether we should go ahead and do it," said Tony.

Once the helium was inside the dewar, a small miracle took place: The reluctant FIRAS mirror transport mechanism finally worked. The mirror mechanism's tiger team, which John Webb headed, had done everything humanly possible to ensure its proper operation, but nobody knew what would happen once it was inside. It was beginning to look as though we had beaten our hardware problems at last. We began talking with the McDonnell Douglas launch crews about a date in June or July 1989, even though several major test hurdles remained.

Among these was the crucial "environmental test": The entire COBE, including its suite of instruments, was placed in a gigantic vacuum chamber at Goddard in order to simulate as closely as possible the inhospitable environment of space. One by one, every complex system and subsystem was checked off on a verification matrix, a formal list required by NASA. Before we could do the environmental test, we had had to pass the vibration test. The most frightening of all to me, it was scheduled for March 1989. A huge electromagnet capable of applying a vibrating force of 35,000 pounds would shake COBE like a mixing machine at a paint store. If a malfunction occurred in the shaker control electronics, the work of fifteen years could be ruined in an instant.

The engineering team was careful, the shaker worked properly, and COBE passed with flying colors.

✳ 15 ✳

Lost in Space

I invent nothing. I rediscover.

—AUGUSTE RODIN

IN EARLY SPRING 1989, headquarters pushed us for a launch date as early as July. We were being watched closely: NASA needed a brilliant success, Len Fisk, NASA's associate administrator for Space Sciences, told the COBE team, not just a mission that might come close; it had been three years now since *Challenger,* and COBE's costs were running a million dollars per month. The science team met on April 14 to discuss this new charter from headquarters as well as COBE's state of readiness, unaware that the night before a calamity had just occurred.

So far, all the COBE tests had taken place with the spacecraft vertically upright. John Sudey, an outspoken senior mechanical engineer, had insisted on testing the instrument with the spacecraft axis horizontal and the FIRAS mirror transport mechanism and calibrator axes vertical in order to check out the troublesome apparatus in that orientation. During the test, which employed specially built equipment to hold the COBE horizontally, the mirrors worked perfectly. But without warning, FIRAS's external calibrator failed during the test.

The movable device was the heart of the instrument, emitting a spectrum within 0.01 percent of a perfect blackbody. It moved in and out of FIRAS's input horn and could be adjusted to match the flux from the sky very closely. Remaining spectral differences between the blackbody and the sky would then be measured with a sensitivity never before possible. Now without gravity to help hold it in place, the calibrator popped out of the horn every time the test engineers inserted it by means of the same electronic commands they would use once COBE was in orbit. Nothing the engineers tried would keep it in place.

"This is a showstopper," John Wolfgang told his engineers that morning.

Without the calibrator, FIRAS would have no reference point, and its measurements would be useless. Over the next few days the engineers considered every possibility: that the flexible cable to the XCAL, as it was called, was too stiff, that a bit of dirt or something else was wedging its mechanism, that the XCAL's kapton skirt had become too stiff in the extreme cold inside the dewar, that its mechanism was misaligned, or that the XCAL was not sliding into place properly inside the sky horn. But nothing worked. The XCAL would not stay put.

Only one solution remained: The engineers would have to examine the XCAL directly. It would take more than a month to pump down the liquid helium to warm the dewar enough to open it. This would totally shoot the schedule and budget, using every penny that remained. Along with several engineers, I went downtown to headquarters armed with a videotape Mike Ryschkewitsch, an instrument engineer, had made of a calibrator arm moving in and out of a test cryostat. By now we were fairly sure the calibrator's cable was too stiff, acting like a spring and pulling it from the horn. The only good news we had for Charlie Pellerin at headquarters was that the engineers believed they would not need to remove the instruments from the flight cryostat, since the XCAL would be accessible as soon as the lid was pulled off. The single ray of hope was that, once the lid was off, we could do it quickly.

"Don't take any risks without letting us know," Pellerin said, granting permission to open the cryostat. He expressed concern that the COBE engineering team was burning out from overwork. The entire project was at risk, along with the tremendous human cost. He reminded me that the *Challenger* launch crew had been working eighty-hour weeks for months before the explosion. "Whatever you do, be careful," he said.

Weeks were required to warm the dewar. During this period Mike Ryschkewitsch devised a fix for the calibrator: The cable would be replaced by three thin ribbons of kapton onto which the electrical circuitry would be printed. After many rehearsals, the repair was successfully undertaken in July. With the dewar opened up like a giant Thermos bottle with its top off, the engineers seized one final opportunity to inspect the DIRBE's mirror for dirt, a constant worry, and found it as clean as when the instrument had been placed in the dewar months earlier. I took a ceremonial final look at the instrument before the cover was replaced.

The scheduled launch date was now November 17, 1989. The XCAL repair had consumed three months and $3 million. Had John Sudey not insisted on the horizontal test for the FIRAS mirror transport mechanism, the XCAL along with its stiff cable would have been sent into space, where it certainly would have failed to operate properly, invalidating all FIRAS data. The COBE team had discussed the potential problem, but never imagined its seriousness. For my part, I had made an unconscious assumption that the XCAL cable, which was like every other cable with which I had ever worked, did not require a special calculation or mathematical proof. We were lucky in our bad luck. Something went wrong before launch, and we had the chance to fix it. Our engineers often said that Murphy's law was optimistic—it says merely that anything that can go wrong will go wrong, whereas the pessimist assumed something would not go wrong until it was too late to fix it.

As the spacecraft was prepared for shipment to California, and the schedule got tight, the science team insisted on completing the final test and convinced Roger Mattson that it was still essential. The spacecraft would be tracked by radar from the moment it was launched and throughout its orbital life. Would COBE's exquisitely sensitive instruments be overly sensitive to radar? Launch-site radar operators had already agreed not to operate their equipment at full power until the spacecraft was miles away; in any event, the instruments would be safely locked in the "off" position during launch. But what about the immensely powerful ground-based radars that tracked every orbiting satellite and space debris bigger than a few inches? They would lock onto COBE with several volts of signal even from a distance of a thousand miles. A fraction picked up by the scientific detectors could be disastrous.

The COBE spacecraft was moved one last time into a large test chamber where it was bombarded by test radar signals. Each instrument passed with flying colors, with only minimal effect on the detector itself. But to my surprise, the spacecraft's Earth scanners that would ascertain that the satellite was pointing away from the planet stopped operating altogether. This was another potential disaster: The COBE could tumble out of control the moment radar locked onto it. The sensors' manufacturer had promised us that the problem should not exist. This was not comforting and, moreover, it was too late to buy new sensors; we would have to fix them ourselves.

Harvey Moseley suggested the solution: Install Faraday cages. Michael Faraday, the nineteenth-century pioneer of the laws of electricity, discovered that electric fields could not exist inside an electrically conductive shell. So COBE engineers constructed a birdcage-like shell with very thin wires around each sensor. If the scheme worked, the sensors' infrared signal would pass between the wires while the radar signal with a longer wavelength would not be able to penetrate the cage.

The Faraday cages did their job on the first test, averting one last, my colleagues and I hoped, potential disaster. The conversion of the spacecraft was now complete. The team working at Goddard and for outside contractors had accomplished a miracle: cutting COBE's weight in half while reconfiguring three of the most sensitive instruments ever built so they would not interfere with each other or the spacecraft, or pick up erroneous signals from Earth and the cosmos.

The COBE was as ready as it would ever be for its date with destiny at Vandenberg Air Force Base.

A THIN GRUEL

While COBE was being reconfigured for a Delta launch, members of the science team, particularly those of us stationed at Goddard, had little time to follow daily developments out in the universe. The information that did reach us from time to time indicated that the picture of the cosmos in the late 1980s was not as simple or pretty as it had been when we had begun thinking about COBE fifteen years earlier. The good news was that the dipole anisotropy, which two groups had detected faintly in the early 1970s, had been confirmed in 1979 by a Princeton high-altitude balloon team led by Dave Wilkinson and reconfirmed several times by other groups, including one led by

George Smoot.[1] Of course, cosmologists anticipated discovery of the dipole all along; the surprise would have been if it had *not* been seen.

Bigger news gripped the astronomy community in 1981 when two teams—one led by Dave at Princeton and the other by Francesco Melchiorri of the University of Florence—announced almost simultaneously that they each had detected with balloon-borne instruments evidence of a cosmic quadrupole.[2] While the dipole is a pattern in the sky with one slightly warm pole and a relatively cool one caused by the Doppler effect of the earth's motion through the cosmic background radiation, a quadrupole pattern consists of two warm poles and two cool ones. If the findings were correct, this would be the first sign of anisotropy in the cosmic microwave background radiation unrelated to the motion of the earth or the Milky Way.

Theorists speculated that several things could create such a pattern: a rotating universe, one region of the universe expanding more rapidly than another, or, most exciting of all, evidence of the long-sought anisotropy in the cosmic background radiation that could indicate how the current universe evolved from the Big Bang. Other experimenters tried to duplicate the results using balloons and measurements taken at the South Pole, but never managed to confirm the Wilkinson and Melchiorri discoveries.

A greater shock to the COBE science team, especially to me since I was in charge of the FIRAS instrument, was an announcement made in early 1987 by a Japanese-American team headed by Paul Richards, my old mentor and friend at Berkeley, and Toshio Matsumoto of Nagoya University. The Berkeley-Nagoya group had launched from the Japanese island of Kyushu a small sounding rocket carrying a spectrometer some 200 miles high. During the few minutes it was able to generate data, the instrument measured the cosmic background radiation at six wavelengths between 0.1 millimeter and 1 millimeter. The results were disquieting, to say the least: that the spectrum of the cosmic background radiation showed an excess intensity as great as 10 percent at certain wavelengths, creating a noticeable bump in the blackbody curve.

The cosmological community buzzed with alarm. Theorists wondered if the excess temperature bump was the modern relic of an early burst of energy that had rocked the universe shortly after the Big Bang. Or worse: Was there something altogether wrong with the standard Big

Bang model itself, as Fred Hoyle and his stubborn band of colleagues still insisted? By the late 1980s Big Bang theorists already were shaken in their ideas about the origin and evolution of the universe because of a succession of unexpected and uninvited new astronomical observations. In 1987 a group of seven astrophysicists who had analyzed the motions of some four hundred galaxies in our region of the universe made a surprising announcement: The Milky Way, along with every nearby galaxy, was streaming at a rate of 600 to 700 kilometers per second toward a point in the sky some 300 million light-years beyond Hydra-Centaurus, a supercluster 70 million light-years away.

Nicknamed the Seven Samurai, the group calculated the mass of the new supercluster, which they dubbed the Great Attractor, as that of thousands upon thousands of galaxies the size of the Milky Way.* Two members of the group, Sandra Faber and Alan Dressler, soon announced that the Great Attractor appeared to be two extremely dense superclusters of galaxies 300 million light-years across. From its gravitational effects, Faber and Dressler calculated the mass of the Great Attractor as about 10,000 trillion (10^{16})[†] times that of the Sun.[3] About the same time, a group of astronomers at the International School for Advanced Studies in Trieste, Italy, discovered a thick concentration of galactic structures in a region of the universe about three times as distant as the Great Attractor.[4]

The large new class of superclusters alarmed theorists, who were unable to explain how such large structures had emerged from what appeared to be a smooth expansion ever since the Big Bang: Where were the seeds of these huge galactic clusters and superclusters? No explanation was forthcoming, and worse news was on the way. In the mid-1980s John Huchra and Margaret Geller, both of the Harvard-Smithsonian Astrophysical Observatory in Cambridge, Massachusetts, began a redshift survey of a narrow band of sky with a 60-inch tele-

---

* The Seven Samurai were Gary Wegner, Dartmouth College; Alan Dressler, the Carnegie Institution; Sandra Faber, University of California at Santa Cruz; David Burstein, Arizona State University; Roger Davies, Kitt Peak National Observatory; Donald Lynden-Bell, Institute of Astronomy, Cambridge, England; and Robert Terlevich, Royal Greenwich Observatory, England.

[†] For large numbers, note that a European billion = 10^{12} and a trillion = 10^{18}, while a U.S. billion is 10^9 and a trillion is 10^{12}.

scope at Mount Hopkins, Arizona. Piecing together data accumulated over several years, they announced they had found the largest coherent structure ever seen in the universe: a system of thousands of galaxies at least half a billion light-years across and only 15 million light-years thick. They christened it the Great Wall.[5] Left with no explanation for gigantic structures such as the Great Attractor and the Great Wall, theorists were stunned.

"These new superclusters we're seeing are like a vise closing in on our theories," said Edwin Turner, a theorist at Princeton University. "We're starting to find that we just don't have enough time to get the universe from an early state to the one that we're seeing now. I don't have the feeling that we're working in an area we understand anymore."[6] Other theorists began admitting they were feeling trapped in an extremely clumpy universe that had started out far too smoothly. Even the miracle of Alan Guth's inflation, along with the many subtheories it had spawned, could not help. Jim Peebles, who by now had become a sort of father confessor to the theoretical community, conceded he was worried about the proliferation of new theories purporting to interpret the inconsistencies.

"Indeed I sometimes have the feeling of taking part in a vaudeville skit: You want a tuck in the waist? We'll take a tuck," Peebles stated about the efforts of theorists to accommodate all the new observational data. "This is a lot of activity to be fed by the thin gruel of theory and negative observational results, with no prediction and experimental verification of the sort that, according to the usual rules of evidence in physics, would lead us to think we are on the right track."[7]

As the theoretical environment deteriorated, Peebles joined forces with Joseph Silk, a leading theorist at Berkeley, in the creation of a statistical analysis of the merits of the five leading theories for the large-scale structure of the universe.* Silk and Peebles analyzed the theoretical models in terms of their fit with thirty-eight kinds of observational data, including the cosmic background radiation, quasars,

---

*The theoretical models Peebles and Silk analyzed were cold dark matter, hot dark matter, cosmic strings, galactic explosions, and baryonic dark matter. Silk was a leading advocate of a baryonic dark matter theory in which compact stellar remnants such as neutron stars and white dwarfs would provide the gravitational means to create large-scale galactic structures.

the huge new galactic structures, small-scale clustering of galaxies, and the internal structures of galaxies. The result of their analysis: no clear winner.

"To enrich, enlighten, even amuse those of our colleagues who are trying to assess the merits of rival cosmogonies, we have begun a model programme of setting up a cosmic book of odds," they wrote in a paper titled "A Cosmic Book of Phenomena," published in *Nature*. The article described the theoretical state of the universe as "dismal," while the number of new observational phenomena was impressive. Unfortunately, none of the theories was consistent with the newly observed data over a wide range. "Therefore, we would not give very high odds that any of these theories is a useful approximation of how galaxies were actually formed," Peebles and Silk concluded.[8]

By now the worldwide cosmology community was as eager as my colleagues and I for COBE to fly. A successful mission, Peebles, Silk, and the other theorists hoped, would bring to an end the growing confusion about how the universe became what it is.

FIRST STEP

The first stage of COBE's long journey from its clean room in Building 7 at Goddard into space occurred on October 4, 1989. Along with the expectations, hopes, and fears of the hundreds of people who had designed and built it, COBE was packed into a specially built container and loaded aboard a special flatbed truck. After a short ride around the Washington Beltway to Andrews Air Force Base, the entire truck and its multimillion-dollar load were driven up a ramp into the belly of a C5 aircraft. The plane flew nonstop to Vandenberg. (By a quirk of timing, the Hubble Space Telescope was flown the same day from California to Florida for its launch, passing COBE en route.) Tony Fragomeni and a big contingent of engineers already had flown out and met COBE at Vandenberg, where they placed it in another clean room to be prepped and tested for launch. I stayed at Goddard to work on final tests for FIRAS.

"You need to get out of here for a while," Ray Weiss told me. "If you don't leave now, you won't get another chance for a long time." I was barely persuaded. There was little more I could do before launch to ensure its success so my wife, Jane, and I flew to Paris for a week. The holiday proved to be a wonderful and necessary respite from the extraordinary pace of the previous three years. Project manager Roger Mattson took a break too, but for a different reason: He had prostate cancer, which had so disabled his nervous system that eventually he could get around only on crutches. It claimed his life around six months later.

The Delta rocket was shipped by truck from Florida where McDonnell Douglas employees had assembled and tested it. Tony was as surprised as I when he saw it for the first time. The rocket was corroded in spots, its aluminum hull pieced together from parts covered with patches. The Goddard engineers began calling the rocket's first stage the "AIDS quilt." Tony claimed unconcern. "When you've been in this business as long as I have, you do the best you can and put all the things that can go wrong out of your mind," he said.

Final tests on COBE proceeded. On the afternoon of October 16, Tony sat in an office near the clean room. As he and another Goddard engineer, John Webb, looked out the window, they saw a telephone pole and its lines quiver. "What the hell's going on?" John asked Tony. The phone rang almost at once. Calling from Maryland, Dennis McCarthy had a television set in his home tuned to the World Series when the 7.1-Richter-scale earthquake struck San Francisco.

"Is COBE okay?" Dennis asked.

Tony said he did not know and would call back. Then he hung up, and said, "Get Mark." However, Mark Flanegan, one of the key COBE test engineers, was not on the site. He and Abigail Harper, another young engineer who was the flight assurance manager and Mark's girlfriend, had taken the afternoon off with Tony's permission. No sooner had they left than a problem developed, and they had turned the instruments off. Irritated, Tony tried to beep Mark, then started calling local motels, to no avail. The engineers began checking out COBE and its precious load of instruments. Tony was particularly concerned about FIRAS, which he regarded as the "world's most sensitive seismometer." If the instrument happened to have been unlocked to run a test, it could have been severely damaged by the earthquake. Tony wanted to ascertain that FIRAS was safe. But that required unlocking it. What if there were an aftershock? John Webb devised a test: He suspended a small weight over a cup of water. Ripples in the water should indicate that the "seismometer" had detected another earthquake.

"Quite scientific," McCarthy declared upon hearing of the technique. But it worked.

But where were Mark and Abbie? Taking an afternoon off just a month before launch was all but unheard of. Late in the day the couple finally appeared. Agitated, Tony asked where they had been. Abbie came over and whispered something in his ear. Tony smiled. They had just gotten married in Santa Barbara. Had they not, Tony recognized, Mark could have been conducting tests on FIRAS when the earthquake hit.

"It would have wiped it out," Tony said. "It would have taken six months to warm up the dewar, repair it, and cool it down again." Fate works in mysterious ways.

In early November COBE made its final trip over land. From its clean room it was loaded onto a spacecraft transportation vehicle for an 11-mile ride across Vandenberg to launch site SLC-2W. The move was made after midnight, when risk of a traffic accident on the base's roads was at a minimum. McDonnell Douglas engineers completed the delicate task of mounting COBE atop the Delta rocket and installing fairings around it. "We're really happy something as significant as COBE is going to be riding this Delta," one of the McDonnell Douglas engineers said. He and his colleagues had feared they would be asked to launch the rocket into orbit as a target for Strategic Defense System ("Star Wars") weaponry.

Once COBE was installed on the rocket, the only way to communicate with the spacecraft and its instrument cargo was via electrical cables. COBE passed its final "aliveness" test on November 9, and its computer procedure rehearsals a few days later. By then the launch had slipped a day to Saturday, November 18, because of a problem with the DIGS, Delta Inertial Guidance System. The date would be the last one available for a while: the following day was ruled out because all NASA communication resources would be reserved for the landing of a space shuttle, and the McDonnell Douglas crew was scheduled to return to Florida for another launch the day after that.

The launch crew stayed up all night conducting last-minute tests. I tried to sleep but could not, and gathered along with the two thousand other invited guests hours before dawn at several designated viewing sites safely removed a mile or two from the launchpad. Mike Hauser stood next to George Smoot at another site. They could not speak as they prepared to witness the spectacle they had been working toward for fifteen years. Other members of the COBE team watched on monitors at Goddard.

Top officials from NASA headquarters were there. As the final seconds of the countdown commenced, Len Fisk stood next to the McDonnell Douglas engineer who was to give the Delta its final command to launch. Fisk had authority to abort the procedure up to the last second if necessary. At approximately 6:34 A.M. the sky lit up and the rocket slowly and, at first, in eerie silence began lifting off the pad. Chuck Bennett, seeing the flash of light, thought at first that the rocket had exploded. Within seconds it was racing faster than the speed of sound, its contrails winding dramatically around a quarter Moon.

"It was a gorgeous, joyful sight," Chuck recalled. Many in the crowd gasped as, a little more than a minute after launch, six solid fuel rockets were ejected at an altitude of more than 10 miles. But not Dennis McCarthy. He had been unable to watch. (See launch illustration and figure 8.)

Within an hour the COBE was a satellite, circling the planet in an almost perfectly circular north-south orbit with an altitude ranging from 899.3 kilometers to 900.5 kilometers that followed the moving dividing line between night and day, making a complete circuit every 103 minutes. Dennis McCarthy was elated.

"What a joyful ride," he said, adding, even though he had not witnessed the first moments of the launch, "That was quite a show."

Nancy remained at Vandenberg for the launch postmortem, thanking the launch crew on behalf of the COBE team. This was not only a polite thing to do, but beneficial as well. Her gracious words meant a lot to the engineers, many of whom, their jobs suddenly and dramatically ended after years of strenuous effort, fell into a postlaunch emotional letdown in the manner of *post coite omnis animalis triste est* ("after coitus every animal is sad").

My colleagues and I on the COBE science team, of course, did not have the luxury of becoming depressed after launch. In a very real sense, our work had just begun. We had to get back to Goddard and find out what kind of data COBE would generate about the beginning of the universe in the hope that the knowledge we would gain would justify its $160 million cost (not counting Civil Service salaries or the rocket) so far. The total cost, counting everything, was probably about $350 or $400 milllion.

COBE DELTA LAUNCH.
COBE was launched southward on Delta rocket no. 189 at 6:34 A.M. Pacific Standard time on November 18, 1989, from the Western Space and Missile Center at Vandenberg Air Force Base in California.

FIRST LIGHT

By the time I arrived back at Goddard, the operations engineers had already received COBE's first telemetry as it flew over Greenland. This informed them that the satellite was alive and well and had sprouted its solar panels and communications antenna as planned. The Delta rocket

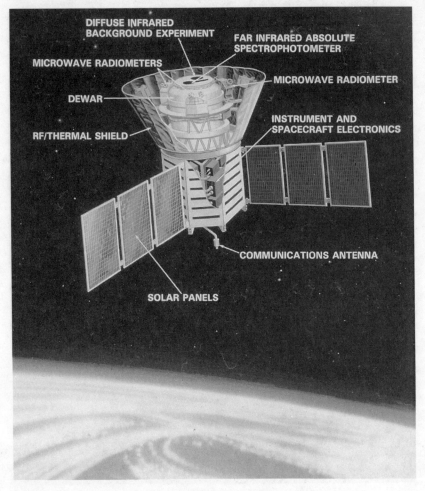

Fig. 8. COBE IN SPACE.
Artist's depiction of how the COBE spacecraft might have looked in orbit with its three solar panels deployed.

had placed COBE in a safe position waiting to be switched on and had opened the vent valves, and now the engineers' first job was to get its attitude control system up and working properly. More complicated than the scientific instruments, the system required some twenty complex equations to describe the way its interlocking controllers worked together; the attitude control system, moreover, had been impossible to test fully on the ground and was a source of worry.

The operations team went through a sequence of commands so long it occupied yards of computer printouts in order to turn on large spinning wheels inside the spacecraft. The COBE began rotating on its axis like a piece of meat on a rotisserie over a barbecue, heating evenly on all sides in the face of the unhindered Sun 900 kilometers above sea level. The next step was easier: starting up the DMR instrument. Since it was outside the dewar, all the engineers needed to do was send a command to turn on the electrical power—almost as easy as pushing a button on a remote control to turn on a television set. Still, nobody dared breathe. The commands went up. Within seconds the DMR sent a signal back to Earth, almost immediately showing a large spike in the data graph as the line of sight passed by the Moon. The engineers cheered. This signal meant that the DMR instrument, at least, had survived the heavy vibrations and stresses of launch as well as the powerful radar signals now tracking the satellite.

Next on the agenda was starting up the FIRAS and DIRBE instruments inside the dewar. Again, the engineers sent up the remote commands and again they received the expected data signals. Unlike those of the DMR, these signals were of absolutely no astronomical significance: The DIRBE and FIRAS were observing the underside of the dewar's lid. The lid could not be removed until the engineers waited for grease and moisture residue from the launch to evaporate or float away into space. Nobody knew exactly how long this would take. But in a practical sense we could not wait too long: We needed to get at least *some* data into our computers in the event that a major problem developed with the satellite or its elaborate electronic communications system. It would be unconscionable to end up with nothing to show for all the effort and expense simply because we had waited to long to pop the lid.

Almost at once our fears about something going wrong were justified.

Almost immediately, the cryostat outshell began to cool down, so quickly we feared the cover might not come off in time. Then, four days after launch I left the scientific control center in Building 7 about 3:00 A.M. and went home to get some sleep. Rick Shafer remained to check FIRAS data and send the instrument another string of commands. At 5:30 A.M. the phone rang beside my bed. Jane heard me say, "You mean we've lost the mission?" I got up immediately, dressed, and went into the kitchen and drank a glass of milk for breakfast. I got in my car and drove back to Goddard. Rick had told me that one of the six gyroscopes stabilizing the satellite had failed. He feared that the loss might cause COBE to tilt precariously one way or the other in its orbit, perhaps facing directly into the Sun, which would be fatal for all three instruments. Alone on empty streets in the predawn November gloom, I thought to myself: If COBE's dead, it's dead, and there's nothing I can do about it. But if it's not dead, we need to work out our strategy very, very carefully.

Arriving at the control center, I breathed a sigh of relief as Rick told me, "Don't worry—it switched over to the backup system and everything's fine." But, Rick said, nobody knows what caused the problem. How did we know that another gyro would not die within a few more days? Or what if a gyro in the backup system failed? As dawn broke, the engineers recommended we switch control of the spacecraft back to the primary system to see if it would operate with one fewer gyro. If it failed, COBE again would switch back automatically to the backup system. The commands were sent up, and it worked, although a bit wobbly. The spacecraft continued to spin properly in a slightly realigned mode that would last for several years (until another gyro would begin to weaken).

Later that day John Wolfgang, who was now running the satellite's operations control, and his team prepared to blow away the dewar's aperture cover. This, I had known for years, would be, aside from the launch itself, the most nervous moment in the entire mission. The dewar lid was held down against springs by a clamp that would be released as its bolts were cut by an explosive charge, similar to firing a rifle with its barrel sealed. The spacecraft had to be tilted back somewhat during the tricky procedure to allow the lid to fly away into space. If the operation did not work as intended, the lid could on a later orbit fly back into the spacecraft or, worse yet, into the instruments.

The operations engineers began the command sequence.

A moment passed. The signal came back: The slight air drag at 900 miles did what we hoped it would do, pulling the lid safely down below COBE. At last the FIRAS and DIRBE instruments, which we hoped would do so much to explain its origin and evolution, were exposed to the cosmos for the first time. No sign of the lid was seen again.

Late that night we turned on the FIRAS for the first time in orbit, almost immediately receiving back its "first light"—the expression astronomers use for the first data reaching a new telescope (today, of course, *any* detecting instrument). Everybody there signed the computer printout, and I took it to hang on the wall of my office. Still, though, we had learned nothing new about the universe: The instrument had not been calibrated in space, and we feared the calibration procedure itself could cause irreparable harm. If its motor failed to work properly, the XCAL could stick in the horn, as it had on the ground, blocking out further input from the universe. We planned to let the instrument collect cosmic background radiation on its own for a couple of weeks before attempting the calibration. In the event the XCAL then damaged the antenna, we would have at least some data to show for our efforts.

Members of the science team monitored COBE around the clock. Until FIRAS was calibrated, however, there were no data for them to interpret. One night after midnight, Ed Cheng and Rich Isaacman watched the satellite's progress on computer screens. They were bored. Cheng had an idea.

"We've got the data pouring in," he said. "We just don't have the calibration yet. We've got the ground-based calibration, though, and it should work fine."

Ed and Rich began working on their monitors. They pulled up the original calibration done in Building 7 at Goddard and undertook a couple of maneuvers aimed at fooling the computer into thinking this was actually a space calibration. Then they started comparing the ground calibration with the raw data that were already flowing in from the spacecraft. After a few minutes, the monitor started flashing.

"Holy shit," Ed and Rich said in unison as the screen displayed an absolutely perfect blackbody curve, the slightly skewed arch that everybody desperately wanted to see.

"It was a beautiful, smooth spectrum," recalled Rich.[9] "I knew right away that it was right. I was almost positive that this was exactly the same thing we would see when we did the space calibration."

He and Ed made a printout of the screen. Ed began plotting points on it by hand. Rich called off numbers from the monitor. Soon Ed hardly had to wait for the numbers. He already knew where the points would fall. By now a small crowd of people from the other instruments was hovering over Cheng's shoulders. Nobody spoke. Cheng plotted the points, and it became more and more apparent that COBE had found a perfect fit to the blackbody curve at all wavelengths.

Rich and Ed recognized at once that the Berkeley-Nagoya results had been wrong. COBE's FIRAS instrument was many times more sensitive. Rich felt he was looking at the hand of creation itself as the earliest moment in the history of universe yet beheld by humans appeared on their graph paper. The whole team was sworn to secrecy. NASA considered the data proprietary at the time, although they eventually would be released to the public. Moreover, the data were far from complete; it would take weeks, even months, to accumulate and fully analyze them all, and they still might be wrong. Although global, the physics grapevine is small and tight. Calls soon started coming in: "Hey, I hear you guys knocked the bump out of the Berkeley-Nagoya stuff." "When are you going to release it?" "Are you sure you got it right?" We didn't tell them; they were just trying to get us to spill the beans.

When Rich and Ed gave me the news the next morning, I was delighted—albeit cautiously. We still had not done the space calibration, so we did not know absolutely that the data were accurate. Moreover, although we believed COBE was operating the way we hoped, many things still could go wrong. I continued to believe somewhat pessimistically that when everything in a space mission worked as intended, it was as much a miracle as a triumph of technology. A week after Ed and Rich's midnight foray into the uncalibrated FIRAS data, we undertook a proper calibration in space with the worrisome XCAL. It entered smoothly and, to everybody's relief, exited the instrument's horn antenna. We hoped to confirm the uncalibrated results within a few days.

In the meantime, however, the engineers were trying to solve another problem: For unknown reasons, the notorious FIRAS mirror transport mechanism was going to the end of its stroke and remaining there during the period COBE flew through the cosmic ray zone over Brazil and the South Atlantic Ocean. If the mirror kept sticking, we would receive almost no good data, and the helium in the dewar would evaporate long before we could accumulate the amount of data necessary for solid scientific results. It would ruin both FIRAS and DIRBE.

There was only one solution: The mirror transport mechanism would have to be shut down every time it came to the cosmic ray zone. None of the engineers knew whether the mechanism would tolerate being turned on and off an estimated three thousand times. Even something as simple as a light bulb failed most often the moment it was turned on, so experience suggested caution. Fortunately, we had a different problem on the ground that led us to devise special hardware we called the MTM Watchdog—or M-dog—whose job was automatically to take control of the mirror transport mechanism whenever the sticking problem occurred. John Yagelowitch developed it.

Once again, bad luck on the ground turned out to be good luck in space, and my own scientific future did not look as shabby as it had when the engineers first discovered the potential disaster awaiting COBE in the cosmic ray zone. I thanked my lucky stars and my engineering friends again and again. Eventually they determined that cosmic rays striking the mirror transport mechanism's photodetector, which indicated when the mirror was at the beginning of a stroke, produced a signal large enough to confuse the mechanism. Still, there was another surprise: COBE discovered Antarctica. Flying over the South Pole on each orbit, the spacecraft received more solar power than anybody had anticipated reflecting back from the southern icecap. We had spent years trying to cut back on the costs of the spacecraft's solar cells, and now there was a possible problem with too much power! The engineers made adjustments, averting the problem.

The FIRAS instrument now began collecting good data from space in earnest. Within several weeks we had collected nine minutes' worth of really solid observations of the cosmic background radiation, primarily from the North Galactic Pole, where there was little interfering galactic dust. As our first serious spectrogram began emerging, it confirmed what Rich and Ed already knew from their midnight graph: The cosmic microwave background radiation had a nearly perfect blackbody spectrum.

In astrophysical observations, measurements usually have a certain margin of variance from the results predicted by theoretical models. But our sixty-seven measured data points matched the theoretical curve precisely—or at least within 1 percent. The Big Bang theory, recently having such a hard time justifying itself, now looked intact again.

BREAKING THE NEWS

The science team optimistically reserved a place for itself at the winter meeting of the American Astronomical Society in Crystal City, Virginia, in January 1990. The team hurried to get ready, and spent the next several weeks struggling to make the computers work properly in order to better quantify our blackbody statement. The FIRAS instrument was designed to match the incoming signals from the cosmos with a signal from our internal reference calibrator. If they matched well, we would get an extremely small signal. Such an approach is called a "null experiment" because if everything has been adjusted right, the signal comes out zero.

According to theory, residual radiation from the Big Bang has a blackbody spectrum. As soon as we turned on the mirror transport mechanism, we realized the theory was correct: Our competitors at Berkeley and Nagoya had measured a bump in the spectrum that simply had to be wrong. Our instrument was a thousand times more sensitive than that of the Berkeley-Nagoya experiment and was not affixed to an overheated rocket. Still, we had to keep our perfect spectrum under wraps because the COBE science team's publications policy stated that nothing was to be revealed until everybody on the team agreed to the content and form of the announcement. Moreover, we easily could have made a terrible mistake.

But the news was hard to keep secret. At Princeton, Dave Wilkinson first saw the FIRAS-generated blackbody spectrum on his computer screen, sent to him by Ed Cheng by means of electronic mail. Many researchers had gathered bits and pieces of the spectrum over a twenty-five-year period, each point on the line representing great expense and hard work. "Even though I had expected it for years, seeing the spectrum in one piece was simply thrilling," Dave recalled. For over a month he managed to keep the news to himself, even over coffee with Jim Peebles and Robert Dicke, whose offices were near his in the physics building. One day Dave could restrain himself no longer and pulled a piece of paper from his pocket and showed it to Peebles.

"Take a look at this," he said.

Peebles, who had never believed the Berkeley-Nagoya results but still did not expect to see a perfect spectrum, was all but speechless. He recalled later that the moment was one he would never forget. Had the Berkeley-Nagoya distortion turned out to be correct, as Wilkinson had started to fear, Peebles realized that he and other theorists would have to get down to work to revise the Big Bang model or create a new theory altogether. Perhaps having more faith in nature as well as in our instrument, I must admit that, like Jim Peebles, I was not very surprised at the perfection of the FIRAS spectrum.

The cosmic background radiation simply dominated the early universe in which there were approximately one billion particles of light for every particle of matter. If the spectrum were not a perfect blackbody, every bit of matter would have to be extremely busy, and I had never been able to imagine how the necessary energy could be available. On the other hand, I was slightly disappointed. After all, there are only so many scientific papers one could write about perfection.

The science team began preparing a manuscript for publication in the *Astrophysical Journal*. Writing it may as well have been a new enterprise as far as I was concerned. A well-known aphorism in scientific and academic circles is "publish or perish," but I had been too involved with COBE's journey through its long bureaucratic, scientific, and technological obstacle course to have time for writing since COBE's proposal days. Attempting to achieve a marriage of minds over detailed scientific data and a complex text among the eighteen men and one woman who now made up the science team was more difficult than anyone had imagined. The task was complicated by the FIRAS's ongoing mirror problems in orbit and our computer software, which never worked entirely right. Rich Isaacman and Ed Cheng labored night and day to make the calibration software function properly for the first time. Everybody worked straight through the Christmas and New Year's holiday, since our presentation was scheduled for January 13, 1990. Almost as difficult was keeping our excitement to ourselves, since we were sworn to secrecy until the manuscript was in the mail to the publisher.[10]

The day of our presentation arrived. We were scheduled to speak after lunch in a large auditorium at the hotel near National Airport just outside Washington. I was somewhat annoyed, thinking that by the time of our presentation—the last session of the meeting—everybody would have left. After a last-minute photocopying session, I was ready to mail the manuscript. Inexplicably, George Smoot insisted on accompanying me to a mailbox across the street, where we dropped the envelope unceremoniously through the slot. Mike Hauser, George, and I were dressed similarly for our presentation in what must have looked like the COBE uniform—blue blazers and khaki pants—even though I was certain we would be speaking to an empty hall.

Our turn finally came. Ironically, the final session was chaired by Geoffrey Burbridge of the University of California who, along with Fred Hoyle, was a staunch opponent of the Big Bang theory. As we walked into the large room, which I learned later could hold as many as two thousand people, I was astonished: The room was filled to overflowing. Nancy Boggess stood up and gave a summary overview of the COBE project.

Then I took the podium. After describing the instrument's principle of operation, I displayed a graph of the spectrum of the cosmic background radiation as revealed by FIRAS. (See figure 9.) "Here is our spectrum," I said. "The little boxes are the points we measured and here is the blackbody curve going through them. As you can see, all our points lie on the curve." The theoretical blackbody curve predicted how the blackbody radiation should look if it truly had originated in the Big Bang.

There was a moment of silence as the other scientists there grasped the meaning of the data curve. Then the audience rose, breaking into spontaneous ovation. Blushing and with perspiration rising on my scalp, I stood there speechless before the huge crowd. It had never occurred to me that so many scientists would be there or that they would think the preliminary FIRAS result was so important. I momentarily feared they were clapping for me, and thought about saying something about the team effort involved.

For me this was a moment of supreme epiphany. The COBE project had been such a part of my life for so long: all the proposals and reports, NASA's bureaucratic jungle, the scientific problems, the engineering challenges of reconfiguring the space craft after the *Challenger* disaster, all the personnel difficulties. I may have begun to take the result for granted by now. Obviously nobody else had.

Such displays of enthusiasm are rare at scientific meetings. I was entirely unprepared for it. Chuck Bennett told me later he had never seen anything like it, before or since. Perhaps the audience felt a mystical connection to the earliest moment in the history of the universe yet seen by humans. Perhaps they knew how difficult the experiment had been and appreciated the sound design of the instrument. But more likely, most members of the audience stood and clapped with relief. After all, the Big Bang was being questioned, its theoretical integrity compromised by the Berkeley-Nagoya team's mysterious bump in the spectrum curve.

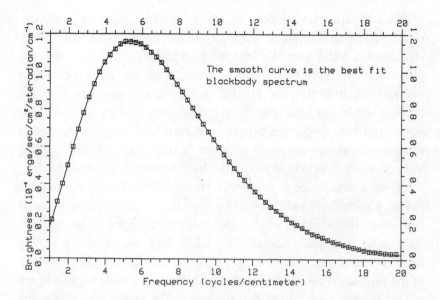

Fig. 9. FIRAS COSMIC MICROWAVE BACKGROUND SPECTRUM.
Spectrum of the cosmic microwave background radiation, based on nine minutes of
FIRAS data. When shown to the American Astronomical Society on January 13, 1990,
it brought a standing ovation. Theoretical prediction is solid line, data points are
boxes.

The FIRAS blackbody spectrum line, perfect in its harmony with
theoretical speculation, confirmed what almost everybody there hoped
and believed was true: that the Big Bang theory really did explain how
the universe began. Our experimental results were essential to their view
of the universe, since any other result would have required a revolution
in cosmology. I hurried through the rest of my talk out of nervousness
and because George and Mike were yet to speak.

Mike spoke about the data coming in from DIRBE, which had not
been processed into a map. At a later meeting he displayed a lovely infrared
picture of the Milky Way compiled by the DIRBE instrument dramatically
depicting the galaxy's plane, its bulging midriff, and its zodiacal light. Mike
and his team used false colors to highlight the map, with red, blue, and
yellow areas representing three infrared wavelengths. This was the first all-
sky picture ever made at the DIRBE wavelengths. Later sky maps produced
by DIRBE caught the public's imagination, appearing in *Popular Science*
magazine among the 100 most significant scientific findings of the year.

One supermarket tabloid, *Weekly World News,* superimposed Grecian columns on the DIRBE map and declared that NASA had discovered heaven. Subsequent DIRBE maps and papers confirmed that the plane of the Milky Way is slightly warped, while its center is not symmetrical, as earlier thought. DIRBE also confirmed other tentative discoveries about our solar system: that the interplanetary dust cloud is not centered on the Sun, as believed, but on a spot several million miles away, due to the gravitational influence of Jupiter and Saturn; and that there is another dust cloud in an orbit near Earth.* (See figure 10.)

George rose to speak after Mike. He told the audience that the differential microwave radiometer was working as intended and that it had seen the dipole, which everybody anticipated. But he also announced that, unfortunately, the DMR had not yet detected any signs of anisotropy in the cosmic background radiation. "Any structure in the early universe will leave an imprint that we should be able to see today," George said. "We're still stuck with the question of where the galaxies come from."

In answer to a question after George's talk, I joked: "We haven't ruled out our own existence yet. But I'm completely mystified as to how the present-day structure exists without having left some signature on the background radiation."

Most of those attending the meeting realized that the FIRAS blackbody results were consistent with the Big Bang model, but the scientific news media emphasized COBE's failure to detect evidence of anisotropy in the cosmic background. An article in the *New York Times* headlined "Spacecraft Sees Few Traces of a Tumultuous Creation" stated the next day that COBE had "revealed an almost too perfect universe."[11] *Discover* magazine later ran a piece entitled "Too Smooth a Universe?" in which George was quoted as saying, "If we don't see

---

*By 1996, the DIRBE data included the best observational limits on the cosmic infrared background to date, although not an absolute detection yet. DIRBE data also provided strong evidence, based on asymmetry in the shape of the starlight from the central galactic bubble (not visible in optical light), that the Milky Way is a barred spiral galaxy, as well as details of the shape and orientation of that bar. Additionally, DIRBE provided the most extensive measurements of the interplanetary dust cloud in the infrared, including the first all-sky polarization survey. These data revealed for the first time sunlight scattered by dust particles in the interplanetary dust bands discovered in thermal emission in IRAS data, particles thought to originate in asteroidal collisions.

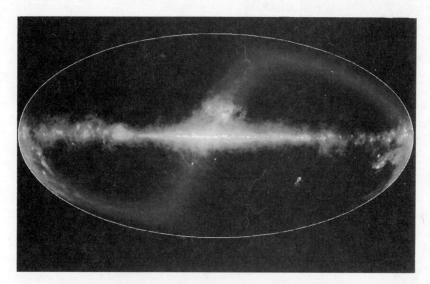

Fig. 10. DIRBE NEAR INFRARED MAP OF THE SKY.
False color map of the sky made by the DIRBE at 1.2, 2.4, and 3.4 micrometers wavelength. The core of the Milky Way galaxy shows clearly. It cannot be seen with visible light because interstellar dust absorbs almost all its light

something, there's something wrong, really fundamentally wrong, with our theories."[12]

I was not deeply concerned. The DMR instrument had been taking data for only a few weeks. As measurements from the instrument accumulated, its sensitivity would increase by many magnitudes. John Wolfgang had pronounced the COBE spacecraft "a sweet bird to fly," and I was becoming more optimistic each day. Barring a major disaster, I fully expected that COBE eventually would begin to see signs of anisotropy, seeds in the early universe of the immense galactic structures that were being detected. Moreover, our preliminary FIRAS data were simply stunning. Bob Wilson, discoverer of the cosmic background radiation with Arno Penzias more than a quarter of a century earlier, was in the audience the day I first displayed COBE's perfect blackbody spectrum.

"It was really, really spectacular," he told me later. "One of the most beautiful scientific results I've ever seen."

✴ 16 ✴

The Private Life
of the Cosmic
Background Radiation

I do not know everything; still many things I understand.

—JOHANN WOLFGANG VON GOETHE

ASPEN, PLAYGROUND OF THE wealthy and celebrated, 7,900 feet high in the Elk Mountains of central Colorado, is not generally recognized as an international center for astronomy or physics. Little known outside the old mining town, a privately funded physics foundation occupies an unpretentious encampment of several one-story buildings amid aspen trees in the fashionable West End. The center is visited often by the stars of cosmology and particle physics. They come to hike, climb, mountain-bike, and raft the Roaring Fork River during the summer and to ski in the winter while, almost incidentally it seems, attending physics conferences. Many of the most famous "phizzies," an endearment coined by the staff, have visited at one time or another: Stephen Hawking, Jim Peebles, Steven Weinberg, Sheldon Glashow, Alan Guth, and Murray Gell-Mann.

A week and a half after the COBE team's announcement of the preliminary FIRAS results, several colleagues and I attended a session at the Aspen Center for Physics to report on our data again. Andy

Lange, a good friend and member of the Berkeley-Nagoya team, spoke about their rocket experiment, and Mark Halpern reported on another rocket experiment undertaken by his group at the University of British Columbia in Vancouver. Headed by Herb Gush, the Canadian team in a stroke of poor timing flew their rocket just a week after the COBE announcement. They, too, had measured the spectrum of the cosmic microwave background radiation with a Michelson interferometer and were in remarkable agreement with COBE, including the radiation's temperature: Gush and his group measured $2.736° \pm 0.017°$ Kelvin, while we had found $2.735 \pm 0.060°$ Kelvin.

This concurrence of temperatures, however, was mainly a matter of good luck since the uncertainties were large in both experiments. The Gush team's instrument was extremely well designed and far more sensitive than ours. But affixed to a sounding rocket, the Canadian detector could observe the cosmos for only five minutes, meaning that the COBE results were more meaningful overall. Gush's group consisted of only a few people. To my mind this made their results little short of astonishing. COBE had become famous and Gush had not, but he and his group had come within days of beating us.

A television crew also visited Aspen. While undertaking a six-part series for public television to be called "The Astronomers," the production team from California had already filmed Lange, Richards, and Matsumoto of the Berkeley-Nagoya group before, during, and after their rocket-borne experiment. Now saying they also wished to film the COBE team, they invited about fifteen of us to a fashionable restaurant one evening and asked us to talk while their cameras rolled. My colleagues and I complied. When the program finally appeared, it focused largely on the Berkeley-Nagoya rocket experiment. The program also suggested that the conflicting results between the two experiments were cause for confusion among cosmologists.

Since it was apparent to everybody in the astrophysics community by the time of the Aspen meeting in early 1990 that the Berkeley-Nagoya results were wrong, the program's thrust was perplexing. The exigencies of television production might have been the determining factor. The crew had spent much time and money filming the Berkeley-Nagoya experiment. When the FIRAS announcement came along, the producers may have viewed it as an unwelcome disruption to their program.

Returning to Goddard, the COBE team held endless meetings aimed at sorting out the millions and millions of bits of data flowing in from the COBE spacecraft. The data analysis was being rendered somewhat problematical by inadequate computer hardware and developmental software still filled with bugs. Worrisome enough, the situation was made even more tense by the fact that headquarters was balking at giving us any more money. Fortunately, Ed Cheng recruited two friends from Princeton: Dale Fixsen, who had been working for Sperry, and Dave Cottingham, who together invented a new mathematical approach that finally made the FIRAS calibration software workable.

The data we were trying to comprehend came by way of a long path after they first were detected by COBE's three instruments. The bits of information were sent directly from the satellite's tape recorder to a NASA ground station at the Wallops Flight Facility on the Delmarva peninsula. Then they were relayed by overland cable to the COBE computer center in Building 7 at Goddard. Software engineers there determined which data were valid and which were not. The data arrived backward from COBE, so the engineers put them in the proper order. When the data at last reached the computers used for scientific analysis, they were unscrambled again and placed in a pattern: so many computer expressions of DIRBE data, then some from FIRAS, bits of data from the spacecraft observatory, then DMR data, and so on. Archives for each experiment were established, with the raw numbers converted into physical units such as voltage, temperature, or electrical current. (See figure 11.)

At this point the process became quite different for each instrument. The DMR data stream consisted of measurements contrasting the difference in brightness of the cosmic background radiation between two parts of the sky, the data so repetitive that each small fluctuation was measured hundreds, even thousands, of times. When sufficient measurements were gathered, a software program produced the map of the sky that best incorporated all the accumulated data. Called a "least squares fitting program," it was arranged to sort out errors occurring in the data-collection process, most of which resulted from effects of the earth's magnetic field on the DMR receivers' Dicke switches. Then the program created the best possible formula to represent the remaining data. When done, the program started over again,

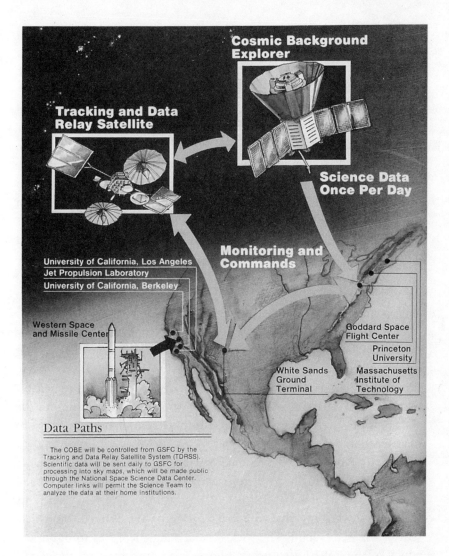

Fig. 11. FLOW OF COBE DATA.
Commands and monitoring are done through the Tracking and Data Relay Satellite System, which replaces NASA's extensive network of ground stations. Scientific data are recorded on the COBE tape recorder and played back daily to the ground station at Wallops Flight Facility.

eliminating additional errors and producing numerous sky maps: at least one for each time period and for each of the six receivers. The science team could then, quite tediously, compare and analyze the maps, correcting again and again for possible errors.

The FIRAS instruments measured interferograms, which could only be interpreted as spectra after being unscrambled by a Fourier transformation. Looking out into the cosmos along the spin axis of the spacecraft, FIRAS sent a set of interferograms back about every 40 seconds, quite different from the virtually constant stream of data from the other two instruments. After the data were sorted out according to the direction the FIRAS faced and how the calibration temperatures and mirror transport mechanisms were set up, they were averaged together. The computer software, programmed to behave like a suspicious human being, compared interferograms that were supposedly identical, then threw out data that looked peculiar owing, possibly, to equipment malfunction or cosmic ray strikes.

The spectra that resulted were then calibrated, meaning adjusted for the temperature of the detectors, calibrators, and antennas. The resulting computerized stew of data produced sky maps for each of the many wavelengths observed by FIRAS. As in the case of all the instruments, a difficult phase of the scientific interpretation involved correcting for the emissions of atoms, ions, molecules, and dust motes heated by starlight to as much as 20° Kelvin throughout the Milky Way.

The DIRBE data set presented yet another package of problems. These related primarily to sorting out millions of bits of information according to where the instrument, consisting of sixteen detectors sweeping the sky together in a helical pattern, was pointed. Most data sent down were sound. The trick in analyzing them was to figure out which of the DIRBE detectors was operating properly since their individual responsivity changed mysteriously over time in a way none of the software experts fully understood. These time-dependent changes, though not large, consumed almost all of the DIRBE team's money and time. Tom Kelsall took charge of the effort to fix the problem, devising a software approach that compared all the stars measured by the instrument with one another as well as with standard stars.

Because many of the stars DIRBE detected are inherently unstable, pulsing on and off over a period of months to years, the problem was uncommonly difficult. Scientific analysis of the DIRBE data was also trickier than interpreting measurements from the two other instruments, owing to interplanetary dust and the Sun. Since these foreground emissions were brighter than the predicted radiation from the distant universe, virtually all of the analytical effort was directed at understanding, correcting, and eliminating the interference.*

Since we had gotten the good news about the early FIRAS result out and with a large amount of accumulated data in hand, we became braver with the FIRAS calibrator. We started moving it in and out of the horn antenna once a month. By summer John Wolfgang's operations engineers and I had enough confidence in the calibrator to swing it in and out every week despite the constant fear it would stick in the horn again. Boldness paid off: After ten months of operation, the dewar ran entirely out of helium on September 21. This was not unexpected. It was simply that nobody knew when it would occur, since the engineers had recommended against running the instrument's helium gauge. They feared its operation would cause the helium to evaporate even faster.

Within the span of a few minutes, FIRAS's temperature soared, making the instrument far too hot to take data anymore. Nobody was unduly concerned. There were more than enough accumulated measurements to occupy the science team for years to come, particularly since, by moving the calibrator in and out of the horn so frequently, the engineers had ensured unusually well-calibrated FIRAS data. That day I was given the honor of sending the command to COBE to turn the FIRAS off, the only direct order I ever sent the spacecraft.

With FIRAS data now residing safely in Goddard's computer banks, the COBE team turned its attention to measurements being taken by the two remaining instruments. Inside the dewar, the DIRBE also warmed up to about 60° Kelvin. But ten of its detectors continued to monitor the cosmos, requiring only recalibration to remain on duty. The DMR, however, was outside the dewar and not affected by the loss of helium. Soon, though, the interpretation of data from the instrument would begin presenting a new set of problems nobody had anticipated and that had little to do with the cosmos.

---

* Scientific interpretation of the DIRBE data was so difficult that as of mid-1996, the COBE science team had yet to issue a paper on the cosmic infrared background radiation. DIRBE team members still hoped to issue such a paper later in the year.

COSMIC SEEDS AT LAST

Measuring and correcting for emissions from the Milky Way was difficult. The emissions resulted from electrons crashing into protons and electrons spiraling around the galaxy's magnetic field lines. Chuck Bennett, George Smoot's deputy at Goddard, spent months devising linear combinations of the sky maps produced by the instrument at different frequencies. He chose coefficients aimed at eliminating the galactic emissions that could obscure measurements of the cosmic background radiation. Chuck also made up a tongue-in-cheek chart depicting the progress of theorists then attempting to set limits on their predictions of temperature fluctuations in the cosmic background radiation.

In 1967 R. K. Sachs and A. W. Wolfe had calculated that the predicted fluctuations would be as large as one-third of a percent. This meant that the early universe would have been extremely inhomogeneous. Over time, according to Chuck's chart, theorists began realizing that large temperature fluctuations were not consistent with the uniform distribution of galactic clusters and quasars in the present universe. They began arriving at ever smaller numbers that would strain the acuity of the DMR analysis. Chuck labored to eliminate galactic noise, and Alan Kogut, a former student of George Smoot's at Berkeley now stationed at Goddard, created elaborate models of possible kinds of instrument faults and software errors. At the same time Ned Wright undertook a scientific analysis of the data sent to his office at UCLA by electronic mail.

By late summer of 1991 Ned was convinced that he saw something significant in the splotchy pictures he had created on his computer screen.* "I had received a bundle of data by E-mail and started studying it at once," Ned recalled. "Gradually I became certain I was seeing a real signal." Working separately from George Smoot and the groups at Goddard and Berkeley, Ned had devised a shortcut method for analyzing the DMR data. "A computer is like a filing cabinet with hundreds, even thousands, of drawers," Ned explained. "If you don't know where to look, you'll never find anything."[1]

Ned undertook numerous simulations of possible kinds of errors in

---

*This section on the discovery of the fluctuations in the cosmic background radiation and its coverage in the press was gathered by co-author John Boslough primarily from interviews with the participants as indicated in the endnotes.

the instrument. Yet the signal persisted. Nothing could explain what he was seeing unless COBE truly had detected temperature fluctuations in the cosmic background radiation. Several weeks after first seeing the fluctuations, Ned could contain himself no longer. He brought a draft manuscript of a paper detailing the exciting finding to an evening meeting of the science team at Nancy Boggess's home. Ned also brought along slides graphically depicting his computer work, proving again Sam Gulkis's observation that he was among the most productive scientists Sam had ever encountered. As far as I knew, Ned had never been wrong on a technical point during the entire COBE project. His work had to be taken seriously. Chuck Bennett thought Ned was one of the brightest people he had ever met. Chuck agreed with me about Ned's result.

"Chuck, you've been working for months now on the DMR data. What do you think of Ned's work?" Ray Weiss asked him.

"I think we've got a signal of anisotropy," Chuck replied.

"I've never heard you say anything like that before," Ray said, commenting on Chuck's notorious scientific conservatism. Ned, in the meantime, urged the science team to release the news about the fluctuations in the cosmic background radiation's temperature.

"The data are good. The signal is there," said Ned. He was concerned about what he felt was inertia on the part of the science team to go public with the news.

The only person not impressed with Ned's work was George. He maintained he still did not believe the data were to be trusted and urged caution. Chuck recalled that George was, in fact, deeply agitated over Ned's presentation. The reason, Chuck believed, simply was that Ned had scooped George and his Berkeley group, which was having difficulty interpreting the DMR data. George had at one time urged that the data be encrypted—or scrambled—in order to keep them away from the prying eyes of the astrophysical community, now deeply interested in COBE's output. Chuck learned of another reason for George's suggestion.

"I don't want Ned to have access to my data," George told Chuck after the meeting.[2]

The encryption suggestion become moot, since George departed for a long-planned trip to Antarctica in November. He planned to make measurements of galactic noise at longer wavelengths than those measured by the differential microwave radiometers aboard COBE. "I couldn't believe that he left for the South Pole when he did," recalled

Chuck. "The COBE data we were analyzing were clearly more important than anything he could hope to turn up there." With George gone for nearly two months, Chuck continued his own difficult analysis of galactic interference with the DMR data. Alan Kogut worked harder than ever trying to turn up possible instrument or software flaws.

As the weeks went by, both Chuck and Alan became convinced more than ever that Ned was right: Without doubt the COBE spacecraft had seen the first signs of anisotropy in the cosmic microwave background radiation. Yet the science team hesitated, aware that the public presentation of science could become distorted. There were good reasons for the worry: On more than one occasion, reporters had told team members that COBE was of little interest; their editors, it seemed, had instructed them to dig for new dirt about NASA and the Hubble Space Telescope. Sometimes scientists were treated like movie stars or heroes, as in the case of Stephen Hawking. Other times the opposite was true. For instance, the press jumped on the cold fusion story when it broke, with some of the best science reporters taking the view that cold fusion was a real phenomenon and that reputable scientists who questioned it were too proud to accept the revolution. In another case, biologist Theresa Imanishi-Kari was accused of publishing questionable data, and her colleague, Nobel winner David Baltimore, spent much of his time defending the work. Propelled by the press, the fight was carried to the halls of Congress. When Baltimore and Imanishi-Kari, their reputations tarnished, were fully vindicated in 1996, little attention was paid.

There were scientific reasons, too, for caution. The history of science records that most experimental results are not as precise as they may have appeared originally. This may explain why even Galileo, Newton, and Darwin were later accused of fudging their data. Ned often spoke of "selective stopping" as a means of causing biased experimental results when the answers began turning out the way you wanted. But now he was sure it would not be wise to stop.

Still analyzing the data at UCLA, Ned was becoming frustrated. George's group at Berkeley was not forthcoming in passing along new DMR data to Ned, even though he had invented a unique algorithm for creating a computerized sky map from the raw material. Ned also started to worry that COBE would be scooped. An MIT group led by COBE team members Steve Meyer and Ed Cheng, and their grad student Lyman Page, apparently had started detecting anisotropy with a balloon-borne detector. Steve and Ed avoided potential conflict by declining to discuss their results with the COBE team; they were as eager as everybody else not to bias the critical COBE data.

Ned sought and was denied permission by the science team to present his draft paper on his analysis of the one-year DMR sky maps at Caltech's Centennial Symposium. He had to content himself with showing an enhanced six-month map instead, stating in his talk that he was sure that the anisotropy signal existed in the data. Not wanting to offend the science team, he discreetly couched his conclusion in negative terms. As a result, his talk received little attention.

By the time George returned from Antarctica, Ned, Chuck, and Alan had finished their draft manuscripts. Ned's paper undertook a cosmological interpretation of the anisotropy being uncovered by COBE. Chuck's work discussed the effect of the galactic emissions on the data. And Alan's analysis attempted to eliminate systematic errors in the data. George visited Goddard in early January 1992 and attended a small meeting in Building 21 with Chuck, Mike Hauser, and me to talk over progress on the data interpretation and to plan a public announcement. After the meeting George asked Chuck to remain behind. George was furious, Chuck recalled.[3]

"This galactic emissions paper is not ready for publication," George said. "None of these papers is ready to be published." Chuck was convinced that George was trying to delay publication in order to give his Berkeley group, which had fallen behind during George's trip to Antarctica, a chance to catch up. "George's outburst was clearly an act of intimidation. He was upset that I had had the temerity to be the lead author on a paper using what he perceived as 'his data,' despite the fact that by common agreement it was COBE data, not any one individual's," Chuck recalled.

Chuck and Ned wanted to go public at the January meeting of the American Astronomical Society. But several members of the team advised caution, since the significance of the discovery warranted detailed verification of every number. Dave Wilkinson, the COBE team skeptic, argued effectively at numerous meetings that he did not believe that Ned and Al had proven that every systematic error in the data was negligible. Dave's worry was that emissions from the earth might be shining over and around the spacecraft's protective shield.

Having missed the AAS session, the science team picked the annual meeting of the American Physical Society, to be held in April near Washington, as the time and place to tell the world about COBE's astonishing discovery. During the following weeks the science team again worked around the clock to write and refine the four papers that would be given at the meeting: the three led by Ned, Chuck, and Al, and a summary paper by George. Mike Hauser, Nancy Boggess, Ray Weiss, George Smoot, and I drove down to headquarters to discuss the logistics of our presentation with Charlie Pellerin and Len Fisk. The APS meeting was scheduled for Monday through Thursday, April 20–23. NASA had already scheduled a talk by another science group earlier in the week, so Charlie and Len decided that we would give our papers at the Thursday session.

"Your announcement will be a great climax to the meeting," Charlie said.

The plan, agreed to by everybody on the science team, was for NASA to issue a series of press releases on various topics. The COBE release would be embargoed for release until noon on April 23; by that time the COBE team would have completed its scientific presentation. A formal press conference would follow, consisting of a talk led by George as the principal investigator on the DMR instrument. Ned, Chuck, and Al would be on hand to answer questions. I wrote the draft of a press release, and worked with Paula Clegett-Haleim of the press office at headquarters and Randee Exler in the Goddard press office to refine it. We circulated drafts to Nancy, Mike, Chuck, Ned, George, and Al. Writing the release was harder than I had imagined, and I began to appreciate the difficulty of making an esoteric scientific topic like ours understandable to the public.

A few days before the meeting, Mike Hauser, who happened to be a member of the APS council on astrophysics that was organizing the event, received a telephone call from another member of the council, a senior scientist at Berkeley. "Are you aware that George Smoot is trying to get himself scheduled to speak on the Monday program?" he asked. Mike was alarmed. The Monday session was for "invited talks." If George were to speak there, it would be in violation of his agreement with the science team and NASA headquarters. Worse, it would scoop the COBE presentation set for three days later.

In the meantime, Nancy Boggess received a phone call from the APS's press director, asking: "Why are you scheduling two press conferences, one on Monday, the other Thursday?" Nancy was baffled. She called George. He told her that he was going to give a scientific talk on Monday.

"George, what's going on?" Nancy asked. "Everybody's agreed with headquarters that we're going to give our talks Thursday."

"Everybody will be gone by then," George replied.

"It doesn't matter," Nancy said. "That's the day we agreed on."[4]

Although she had been the person responsible for securing his position on the COBE team, Nancy was beginning to wonder whether she had made a serious mistake. Nancy talked to Mike Hauser about her conversation with George. Mike acted quickly. After several phone calls to other members of the astrophysics council, he was given assurances that George's Monday talk would be eliminated. The COBE data would be presented on Thursday as originally planned.

"At some level George had started saying in the weeks and months leading up to the APS meeting, 'I don't have to be cooperative with the science working group since they're not being cooperative with me,' even though this was clearly not the case," Mike recalled.[5]

In the meantime, George asked Susan Adams, a COBE team assistant, to make numerous copies of several computer-generated pictures representing the DMR data. Neither the science team nor NASA had approved the pictures for distribution; nobody understood why George wanted them. Susan reported George's request to Nancy, who instructed Susan to bring the pictures to her. When he learned what had happened, George was furious, Nancy recalled. She organized a teleconference with George and Chuck Bennett in an attempt to smooth things over. George refused to participate and returned to Berkeley.[6]

"George apparently was doing everything he could to disassociate himself from the science team and NASA," Chuck believed. Emotions clearly were close to breaking point. It would take little to trigger an explosion.

Two days before the COBE team's scheduled presentation at the APS meeting, a journalist with the Associated Press in Los Angeles called me, asking to read a story he had written about the discovery of anisotropy. Where had he gotten the story? I asked. He told me he had learned about it from a Lawrence Berkeley Laboratory press release. I was flabbergasted. The COBE publication agreement clearly called for full science team approval of all papers, including press releases. George had signed the policy agreement and now had broken it twice: by attempting to schedule his own APS talk and by going to his Berkeley press office with the news. Listening politely, I corrected minor errors. I did not mention that Berkeley did not have the right to issue the press release upon which the article was based.

My reaction might have been different had I known then that the Berkeley release, written by Jeffrey Kahn, mentioned NASA only in passing and did not cite a single member of the COBE science working group other than George.[7] The article, I noted only incidentally, was quite favorable and clearly excited about COBE's remarkable discovery. It included a quote from Stephen Hawking that finding anisotropy in the cosmic microwave radiation was "the scientific discovery of the century, if not all time."*

The day before the APS presentation, Charlie Pellerin and Dan Weedman from Headquarters came out to Goddard to hear George give a dry run of his talk. The consensus among those listening was that George's presentation was disorganized and overly technical. Ray Weiss felt what he described as a "new strangeness in George."[8] Urging George to make a more understandable presentation, we discussed a number of ideas he might try. "Why don't you say that we have discovered the largest and oldest things in the entire universe?" I suggested. I was pretty worried; George looked tired after all the late nights, and I feared he would not be ready.

Charlie Pellerin was more blunt: "Don't fuck up, George. This is front-page stuff—above the fold."

George did not. The COBE team met early the next morning before the meeting at the Ramada Renaissance Tech World Hotel in Alexandria, Virginia. When George stood up to give his scientific presentation for APS members, he seemed like a different person than the day before. His presentation was excellent, full of confidence and real pizzazz.

"We have a quadrupole," George said, smiling to the audience. He explained that the quadrupole was only a part of the irregularities detected by COBE. "These irregularities, unlike the dipole, arise not from our motion in space but from the cosmos itself. They are the oldest and largest known structures in the entire universe." The most dramatic moment of George's talk came at the end when he presented a viewgraph of a map showing the temperature fluctuations in the cosmic background radiation discovered by COBE. After Ned, Chuck, and Alan spoke, the small audience of scientists applauded politely, then asked a few questions.[9]

---

*Hawking later confirmed the hyperbolic remark, which appears on the jacket of George Smoot's book *Wrinkles in Time* (William Morrow, 1993), to Michael Rowan-Robinson, according to his book *Ripples in the Cosmos.* The quote first appeared in the Associated Press story and was repeated worldwide.

The press conference that followed was another story. As we walked into the brightly lit room, my colleagues and I were dazzled. The huge room was filled with cameras, bright television lights, microphones, and, I estimated, more than 100 reporters. The Associated Press wire story, based on the Berkeley press release, had brought them there. As the science team learned later, the press release had been distributed ahead of time, not only to the AP but to a handful of other prominent media outlets, including the *New York Times* and the *Washington Post.* George, Ned, Chuck, and Alan stepped up on the podium.

"We have observed the oldest and largest structures ever seen in the early universe," George began. "These are the primordial seeds of modern-day structures such as galaxies, clusters of galaxies, and so on. Not only that, but they represent huge ripples in the fabric of space-time left from the creation period."[10]

Many questions followed, focusing mainly on the size and fundamental significance of the result. George explained that the structures, manifested as minuscule temperature fluctuations in the cosmic background radiation stretched across the universe for millions upon millions of light-years.

"These small temperature variations spread across the universe are the imprints of tiny ripples in the fabric of space-time by the primeval explosion process," he said. "Over billions of years, the smaller of these ripples have grown into galaxies, clusters of galaxies, and the great voids of space."

In response to a question, George struggled to explain the fundamental nature of the temperature fluctuations. He tried several comparisons, finally stating: "If you're religious, it's like looking at God."[11] Hearing the remark, several members of the COBE team were deeply concerned. None of us ever had discussed our religious views in the context of our scientific work. Most of us saw ourselves as scientists, not theologians, and were wary of having the data from COBE construed as supporting a religious view of nature or, even worse, bolstering one religion over another.

George's single remark, as I had feared, received more media attention than anything else reported about the entire COBE project. He told me later that he was trying to follow Charlie Pellerin's advice and make his presentation more human. But the remark, coupled with the AP story based on the Berkeley press release, converted George into a media star

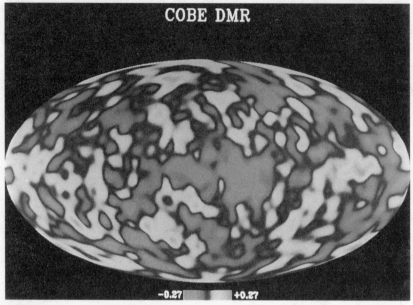

Fig. 12. COBE DMR ONE-YEAR MAP SHOWING ANISOTROPY.
Map of the anisotropy (hot and cold spots) of the cosmic microwave background radiation, made by DMR with one year of data. This image was front-page news around the world in April 1992, and prompted Stephen Hawking to say that this was "the greatest discovery of the century, if not of all time." About half of the spots on the map are due to measurement uncertainties. Statistical analysis shows that there is a cosmic signal as well.

almost at once. With the probable exception of Stephen Hawking, George became momentarily the most famous scientist on Earth.

News of the discovery of the variations in the cosmic background radiation featuring George prominently appeared the next day on the front pages of the *New York Times,* which described him as the leader of the COBE team; the *Washington Post,* making the same mistake; the *Los Angeles Times;* and virtually every other major newspaper in the world. An article in the *Wall Street Journal,* citing an interview with George, stated "he had suspected since August that his space-borne instrument, carried on NASA's Cosmic Microwave Background satellite, had found the crucial evidence of fluctuations of intensity in radiation."[12] (See figure 12.)

George appeared on ABC's *Nightline* with Ted Koppel, explaining the significance of the COBE discovery. Subsequent profiles of George appeared in the *Boston Globe Magazine;* the *Washington Post;* the *New York Times,* which headlined its article "In the Glow of a Cosmic Discovery, A Physicist Ponders God and Fame"; and elsewhere. An article in *Scientific American* compared George to Carlo Rubbia who, the magazine stated, "is said to have won the 1984 Nobel Prize in part by relentlessly whipping a team of physicists into finding a particle confirming electroweak theory."[13] The Canadian magazine *Maclean's* entitled an interview with George "The Man Who Has the Key," and *People* magazine included George among its twenty-five most interesting people of the year, an honor generally reserved for the durably famous.

Neither George's remark about God nor the AP Story was entirely responsible for the media excitement. Scientists, too, apparently believed that the COBE discoveries touched a universal hunger for conscious comprehension, if only metaphorically, of the mystery of the universe's origin and ultimately the origin of our species. Echoing Hawking's remark, Joel Primack, a physicist at the University of California at Santa Cruz, stated that the discovery, if fully confirmed, would be "one of the major discoveries of the century. In fact, it's one of the major discoveries of science."[14]

Michael Turner, a University of Chicago physicist, said, "The significance of this cannot be overstated. They have found the Holy Grail of cosmology." Hawking elaborated further on his now famous remark in London's *Daily Mail:* "It is a discovery of equal importance to the discovery that the universe is expanding or the original discovery of the background radiation. It will probably earn those who made it the Nobel Prize." Another British cosmologist, George Efstathiou, put the COBE discovery in another perspective: "Only the marriage of Princess Di generated equal media interest."[15]

COBE team members were surprised to learn two months later from a profile on George in *Scientific American* that he had sold a book about "his discovery." Tentatively entitled *Wrinkles in Time,* it was to be published by William Morrow & Company along with a number of foreign publishers. George declined in the article to reveal the size of his advance, but intimated that it was very large: "There is enough of a problem with me getting all this press. I don't want people thinking I'm rich, too."[16]

Some six months later, details of George's book deal emerged. John Brockman, the world's leading literary agent for science authors, learned about the COBE discovery from an article he saw in the *Inter-*

national Herald Tribune at Narita Airport near Tokyo, Japan. George was featured, and Brockman got on the telephone to him at once.

"Hey, look, something big is happening in the universe," Brockman reportedly said when he reached George. "What's in it for me?"[17]

A few weeks earlier, Brockman had received a book proposal on the COBE project from the science writer Keay Davidson, a reporter with the *Herald Examiner* in San Francisco. Brockman urged George to co-author a book with Davidson on COBE. George agreed and, with Davidson's help, wrote a new proposal and faxed it to Brockman's New York office. Forty-eight hours later Brockman's agency had a book proposal in the hands of some sixty publishers in more than a dozen countries. Within a few days Brockman secured a series of book contracts in New York and other countries that made science publishing history: approximately $2 million in total advances, reportedly to be divided 70/30 percent between George and Keay.

From the experience of those involved in the discovery of the cosmic background radiation nearly thirty years earlier—Alpher, Penzias, Dicke and the others—science team members were deeply aware of the corrosive effects of a public fight over scientific credit. The COBE publication policy was designed specifically to avoid such disputes. COBE team members were nonplussed, disagreeing over what to do about George's repeated violations of the team's publication policy. His actions were an extremely serious matter in the minds of many team members. As apparent from the convoluted story of the discovery of the cosmic background radiation, due scientific credit was more precious than gold. George cheerfully admitted that he had "clearly violated" the COBE team's publication agreement. But he apparently believed this violation was of little significance, since he believed that the science team was intent on "hoarding credit for itself."

"They always spent too much time partitioning out credit on their scientific papers," George said. "I remember one meeting where they spent a long time discussing whose name would appear seventh on a long list of authors." In light of the acrimony that developed between him and most of the science team following the announcement of the DMR results, George stated later that he "probably should have held off" when Jeffrey Kahn of the LBL press office came to him and said, "We've got to have a press release."[18]

Several team members stated bluntly that they believed George had betrayed them. Ned Wright had done the original analysis of the DMR data when other team members including George and his Berkeley group struggled to interpret the data; he thought that George orchestrated the LBL press release in order to receive maximum exposure in the press for himself.

"The Berkeley press release had an extremely negative result in Los Angeles, where several of us work," Ned recalled. "Sam Gulkis, Mike Janssen, and I were virtually ignored in the *Los Angeles Times,* even though we had been deeply involved in working out the DMR data."[19]

Sam Gulkis had worried about George's attitude all along. "The science working group generally worked very well together with the exception of George," Sam said. "He was never a team player. When we started to get the DMR data he showed his real stripes by doing everything he could to grab all the credit for himself. Unfortunately, as I've seen over and over again, a bad apple always seems to turn up on these big projects."[20]

Nancy Boggess, who had been COBE's and George's guardian angel at NASA headquarters from the very beginning, said she "felt deeply betrayed" by George. "I was horrified at the DMR press conference," Nancy recalled. "George put on a real show. Everything was 'I this' and 'I that.' He didn't even bother to mention Ned Wright, Chuck Bennett, or Al Kogut, who had done all the work." Nancy was particularly upset that George stated that NASA had selected him as the principal investigator on the DMR because of his previous work on the cosmic background radiation.

"That was a lie. And George knew it," Nancy said. "NASA wanted Dave Wilkinson all along. But I knew that Dave would never do it, so I ended up picking George almost by default."

NASA headquarters was as concerned as the science team over the DMR press conference and its aftermath. A press officer phoned Nancy afterward: "How could George do this to us?" he asked. Charlie Pellerin put the same question directly to George, who refused to answer.[21] Some NASA officials contemplated legal action against George and the Lawrence Berkeley Laboratory, but decided against it. "We really didn't want to get in a fight in the press with George Smoot and Berkeley," one official said. Headquarters eventually came to the conclusion that it was best to leave the matter in the hands of the science working group.

Such a course of action presented a real problem, since the COBE publications policy did not contain sanctions for violations. Some members discussed the possibility of asking George to resign from COBE. Ray Weiss told Chuck Bennett one day when they both were attending an astrophysics conference at MIT that he thought George should be removed from the project.

Ray had nurtured and protected George over the years, once even talking him into remaining on the COBE project when George had become discouraged and wanted to quit. Ray "felt a betrayal of a twenty-year trust built up by having worked together through all of the vicissitudes of the project." But Ray also had a distaste for what he believed was the worst that scientists and intellectuals could do to each other when a fight over credit became public. Such a fight, Ray believed, was not only demeaning for the participants, but would not be understood by outsiders.

"There is no question that I waffled and was irresolute, but I could not come up with a satisfying solution, especially after talking with the administrators of Lawrence Berkeley who actually supported George," Ray stated later. "My only comfort was the belief that George's behavior was so outrageous and transparently self-serving that he carried the seeds of his self-destruction. I really got angry with George after his arrogant campaign to garner the Nobel Prize by self-promotion, witnessed by my daughter in Australia, where George was promoting his book and image. The chutzpah of George is unbelievable."[22] Ray met later with George and Chuck at Goddard.

"Chuck, do you want to take over George's job?" asked Ray, who as chairman of the COBE science team had the authority to recommend a change in the position of principal investigator. George sat nervously waiting for Chuck's reply. Pondering the question for several minutes, Chuck at last said no. He believed that if he took the job, efforts to bring about a truce between George and the rest of the science team would be hindered. Moreover, he feared that his replacement of George would appear to the outside world like sour grapes on the part of the COBE team over George's celebrity and book contracts. If removed from the team, Chuck worried, George could end up looking like a martyr crucified on a cross of science team bitterness.[23]

Opinions of team members remained divided. In a pure democracy such as the COBE team, this meant that little action would be taken.

Following intense pressure from Ray and Mike Hauser, George at last agreed to write a letter of apology to the science team. In September, after months of negotiations and after several draft versions, George sent a four-page letter to the science team. While much of the letter was devoted to recounting the COBE project, George acknowledged that he had broken his agreement and he apologized to the team. The portions cited are verbatim:

> Dear Colleagues,
> After 18 years of effort and vision we have finally reached a major milestone that was our goal and the purpose of the DMR instrument. This important step could not have been accomplished without the long term cooperative team effort. Unfortunately, during the final stages several things occurred which strained the team and working relationships. This letter is both to apologize for the major publication's policy breach—the LBL press release—and to address the approach that I intend to pursue as my part in maintaining the cooperative team effort. . . .
>   The publications policy reflects a careful balance of the competing needs of responsibility, credit, orderly process, and freedom for the science team and the additional participants. Even though there have been breaches in the policy by myself and possibly others, it is an excellent policy and I fully support it and will abide by it until the science team declares it ended. I will make amends for my own errors and hope this is reciprocal. Nancy is the coordinator for the implementation of this policy and I will work with her as needed. In order to achieve adequate public knowledge of the contributions of the various team members, I will work with the team to ensure their professional recognition in meetings and publications with the guidance of the SWG. I will also support efforts to clarify these roles to journalists and other media representatives. . . .
>   It is clear that the press reports of the DMR results have been unbalanced in coverage and giving credit for the work of the whole team. This has happened in spite of both spoken and written declarations of the team effort. I have called and sent letters to the authors of articles pointing out that this was a team effort and that there were four papers by four different

authors submitted at once. I have also made a major effort to be sure that NASA and Goddard in particular receive credit. I have arranged and encouraged the various Goddard engineers to be interviewed and be featured in articles. It is clear that very many people feel that the press coverage was unfair in particular neglecting the engineers and other technical people working on COBE.

I do apologize for disruption caused in the team and in the imbalance in the credit and coverage allocated to people. My arguments and work to ensure that various members of the DMR got credit for their work seemed to make relations worse rather than better. I have been strict about upholding the COBE Publication Policy on the release of data, making it a grievous error on my part in not considering the LBL press release as a COBE publication. I considered the NASA and LBL press releases institutional rather than COBE driven. The COBE Publication Policy does include press releases and I apologize for this judgment error. My greatest regret is the net result became a cause of disunity in the COBE team during what would otherwise be a time of great celebration. . . .

George wrote further, promising to stay in regular contact with Ray Weiss, Nancy Boggess, and me in order to prevent future misunderstandings. He also promised to work with Chuck Bennett on "a basis of respect and trust," and noted meetings he and Chuck had had with Ray Weiss and Mike Hauser in order to determine who would be responsible for what area. The letter concluded:

I vow to resolve any misunderstandings I have with individuals on the team in an amicable and private forum. I will take complaints and concerns promptly to the people who can clarify them or do something about them. There is still much work left to be done on all aspects of the mission. This will go most effectively and enjoyably when we all are cooperating. I will be doing my part to see that this happens and we make a smooth transition to delivered products.

Sincerely yours
George Smoot[24]

Nancy Boggess, Mike Hauser, and other team members believed George's apology was halfhearted. In any event, nothing further was done, either by George or the science team. Over time, tempers began to cool, although it appeared that the schism between George and the science team would never be fully repaired.

VERIFICATION

George had brought COBE worldwide publicity, which we might not otherwise have received. In the meantime, however, I had a nagging worry about something else: the validity of the DMR result that had caused such a sensation worldwide. Although Ned, Chuck, and Al had worked hard to verify the signal and found no chinks in the armor, I worried along with Dave Wilkinson, still concerned about "Earth shine," that we had missed a subtle signal somewhere in the universe or in our equipment. As it fortunately happened, other groups also were looking for anisotropy. The Cheng-Meyer-Lyman Page team at MIT had launched in a high-altitude balloon the original bolometers suggested by Ray Weiss for COBE, but which we had decided would not work unless they were colder than we had planned. The instrument concept was close to the design we had removed from the COBE back in 1976, and a very sweet success all around.

The MIT group's detectors, chilled to $0.25°$ Kelvin for even greater sensitivity than COBE's DMR receivers, mapped a large part of the northern sky with a far infrared radiometer at several wavelengths. Cheng and Meyer used the data from the shortest wavelength channels to measure the atmospheric emissions that plague balloon experiments. Then they adjusted the data at longer wavelengths to remove noise from the atmosphere. The resulting maps were similar to the DMR maps, although not all of the temperature fluctuations across the sky were in precisely the same places.

Ken Ganga, a Princeton graduate student, derived a mathematical correlation between the MIT and the DMR maps that, given the measurement uncertainties involved, showed that the MIT and COBE DMR data were in reasonably close agreement.

We had confirmation.

✦ 17 ✦

The Cosmos
After COBE

Success is never final.

—WINSTON CHURCHILL

THE DISCOVERY OF anisotropy in the cosmic background radiation was waiting to happen.

What if NASA had decided not to fund COBE? Or what if the DMR instruments had not been improved sufficiently through passive cooling to improve sensitivity enough to detect the temperature fluctuations in the cosmic background radiation? Other groups were working on experiments aimed at detecting the anisotropy, even though it was widely known that COBE was on the way. Had COBE not been funded and in the building stage, even more scientists would have started working on the anisotropy problem. What if the COBE launch had been backed up another six months? What would have happened had Ned Wright never been invited to join the COBE team? The difficult data interpretation could have been delayed for months.

By the time of our announcement in April 1992, theorists Kris Górski, Sasha Kashlinsky, and others had calculated with great precision the size of the fluctuations that experimentalists could expect to find. The fluctuations just had to be there, or we would not be able to explain the motion of the galaxies. Scientists around the world raced to take up the challenge. Along with the one of the MIT group, balloon experiments were under development by COBE team leader Phil Lubin at the University of California at Santa Barbara as well as by other groups. All had sufficient sensitivity to measure the cosmic blobs that turned up in our DMR data. Even a ground-based instrument, built by a British group and operating at relatively long wavelengths, at Tenerife in the Canary Islands, confirmed some of the details in the COBE maps.

Their job was unusually difficult because the Atlantic weather creates patterns in the air that can produce signals similar to cosmic fluctuations. It took the English scientists years to eliminate this atmospheric noise as well as signals from the Milky Way. None of these other experiments, of course, had the opportunity to use instruments aboard an orbiting satellite above the earth's obscuring atmosphere to measure the entire sky with a single instrument. But the point is that many of these scientists, had they not encountered difficulties and delays, could have done the job.

In particular, the MIT group, which ironically included two members of the COBE team, easily could have been the first to announce the discovery of the famous temperature fluctuations had our announcement been delayed even a few months. And COBE, although producing the most convincing whole-sky picture, would have been left with the backup role of confirming *their* discovery. Such speculation on what might have been is reminiscent of the lingering controversy over the hydrogen bomb developed in the Soviet Union during the late 1940s and 1950s. Soviet scientists doubtlessly were helped greatly by classified details of bomb construction illegally conveyed by spies within the U.S. nuclear program.

On the other hand, the most significant design detail of all was that, like the instruments aboard the COBE spacecraft, the bomb could be built. There was little question that, given determination bred of cold war hostility, the Soviets would make their own bomb eventually. The information they received from the spies merely saved them time and effort. Such was the case with the COBE project, which, owing to its cost and difficulty, was not in a position to be victimized by spies. Everybody in the cosmology community knew how and why COBE was being built. But nobody outside NASA had the means of duplicating it (although the Soviet *Relikt* scientists had tried).

COSMIC THEORIES AND FOSSILS

The burst of scientific activity following the DMR announcement was astonishing. In 1994 David Leisawitz, a COBE team member at the National Space Science Data Center, counted nearly nine hundred professional articles that were based on or related directly to measurements taken by the three instruments aboard the COBE spacecraft. Many cited the nearly perfect spectrum detected by the FIRAS instrument, but most cited the DMR measurements of anisotropy in the cosmic background radiation. Back in 1967 the theorists R. K. Sachs and A. M. Wolfe had stated that detection of the anisotropy, when it finally came, actually would be a direct measurement of the gravitational potential field at the moment of decoupling in the early universe.

This decoupling, as we've seen, occurred when matter became neutral and free to move about the universe, accelerating under the influence of gravitation. Before the decoupling phase, matter was ionized and constantly in collision with photons of the cosmic background radiation. Before COBE, cosmologists were mystified as to how galaxies and superclusters of galaxies could have evolved from what must have been a homogeneous primordial soup of matter and energy. Prior to announcement of the DMR results, theorists around the world had devised computer models of the evolution of the universe. But they could only estimate the numbers for the universe's initial conditions, then hope to guess correctly how to configure their models of the universe to produce galaxies and galactic clusters.

If the theory of Sachs and Wolfe were right, COBE had given theorists the numbers they lacked for the universe's initial conditions. COBE's extremely sensitive measurements of the cosmic background radiation's detected hot and cold spots differing from the average temperature of the sky by only about thirty millionths of a degree Kelvin. But this would be sufficient for cosmologists, if they were very clever in devising their theoretical models, to calculate the evolution of the early universe. Now, without making arbitrary assumptions, they could begin to quantify the actual steps in the creation of the stars (and planets), galaxies, and clusters we know exist today.

Major mysteries persisted after COBE. Among the most perplexing of the unanswered questions is, What makes up most of the matter in the universe? The many surveys of the sky detect only luminous stars.

Yet analyses of the motion of the visible galaxies and galactic clusters based on Newton's law of gravitation indicate matter exists that is not visible to the large array of powerful instruments astronomers have at their disposal today. According to the most popular model today, this "dark matter" consists of a mixture of "hot" and "cold" matter, depending on how fast the hypothetical particles of matter happen to be traveling.* The dark matter has remained elusive, and continues to inspire many searches, measurements, and theoretical models.

Beginning in the 1960s, Vera Rubin of the Carnegie Institution's Department of Terrestrial Magnetism in Washington, D.C., studied the rotation curves of several galaxies, including the Milky Way. She found that each one contained a vast amount of nonluminous matter. This material occurred in a galaxy's outermost reaches, where it vastly exceeded the mass of the galaxy's luminous center. According to Rubin's study, the density of matter falls off from the center roughly as $1/r^2$, where r is the distance from the center of the galaxy. Such a calculation yields a very large total galactic mass.

Another method developed to detect and measure dark matter analyzes the deflection of light from a more distant galaxy around a nearer galaxy or cluster containing large amounts of the mysterious matter. A quick calculation of this "gravitational lensing" based on general relativity reveals the mass of the intervening galaxy or cluster. This method confirms Rubin's method—the masses are large.

According to the Big Bang model, there should be a 77:23 ratio of hydrogen to helium by mass throughout the universe. Since these are the numbers astronomers have measured in stars, the Big Bang model has proved accurate in its prediction of the ratio of the universe's most common forms of matter. However, when theorists first calculated these numbers, they realized there could have been only a small percentage of matter such as helium and hydrogen in the early universe. There can not be enough to account for the total measured masses of galaxies.

---

* Floyd Stecker, a physicist at Goddard, published an article in 1973 with the earliest prediction that the dark matter should consist of just such a mixture of fast and slow particles.[1] The COBE paper by Ned Wright et al., which analyzed the theoretical implications of the DMR findings, failed to mention Floyd's work, even though he went to much trouble to point out his prediction to the COBE team. Understandably, Floyd was keenly disappointed with our error, and I apologized.

Gravitational lensing entices astrophysicists for the simple reason that only a few percent of the galactic masses measured with the lenses appear to come from ordinary matter, a signal that something fairly exotic has occurred: For one thing, at least 90 percent, maybe as much as 99 percent, of all the matter in the universe is unlike anything ever seen on Earth, even at one of the great particle accelerator laboratories such as the one at Fermilab in Batavia, Illinois, or CERN* in Geneva, Switzerland.

More important as far as the COBE discovery was concerned, this mysterious form of matter was a major motivating factor for Alan Guth when he crafted the idea for an inflationary impetus to the Big Bang. Refining inflation, theorists realized that fluctuations in the cosmic background radiation, if they existed, could provide a key to understanding dark matter. Nobody knew exactly what process during the Big Bang might have caused the temperature variations in the background radiation. But it seemed to Guth and the other inflationary theorists a reasonable guess that the fluctuations must have resulted from the quantum mechanical processes that took place in the very early universe. Hence, the anisotropy in the cosmic background radiation related in a profound way to fundamental properties of matter at high densities and temperatures.

Of course, during the 1980s there was no observational evidence of anisotropy. What if no fluctuations were found? Jim Peebles and other leading cosmological theorists began fearing they would need to begin looking at the origin and evolution of the universe in a new light, perhaps abandoning the Big Bang theory for another. Particle physicists were just as worried. If the temperature fluctuations were not seen, they would never be able to test several varieties of the "theory of everything" they are so eagerly developing. Such a theory would unite in a single statement all the interactions of matter and forces, including gravitation, at the most fundamental level of nature. Particle theorists such as Murray Gell-Mann and Steven Weinberg believed that an instant when all matter and all forces were united occurred during the early moments of the universe. Fluctuations in the cosmic background radiation would provide an important clue to that instant of unifica-

---

*CERN is a French acronym for European Council for Nuclear Research, which has changed its official name to European Organization for Particle Physics in order to avoid the stigma attached to the word *nuclear* in Europe.

tion, perhaps even the only quantitative confirmation for the theory of combined forces.

The spectacular enthusiasm for COBE's discovery of anisotropy in the cosmic background radiation was a direct measure of the gasp of relief collectively expelled by the theoretical community as it realized, after years of frustration, that it probably had been right all along. Cosmologists recognized that the anisotropy was a modern trace not only of the Big Bang but of the mysterious and invisible dark matter too. Aside from the helium-to-hydrogen ratio and the blackbody nature of the cosmic microwave background radiation, the anisotropy was the earliest measurable cosmic fossil, and COBE had found it perfectly preserved, buried exactly where it was supposed to be in the cosmic strata.

"Although the COBE results were marginal, they struck a blow for inflation," said Jim Peebles, one of the few who had been in on the cosmic background radiation from the very beginning. "And if inflation is wrong, then God missed a good trick. But, of course, we've come across a lot of other good tricks that nature has decided not to use. So we could still be wrong, although I doubt it. The COBE results could be wrong too, but I wouldn't bet even money on it."[2]

THE FINAL REPORT

In December 1993 the COBE spacecraft was shut down, its scientific mission accomplished. By then, the satellite had started to disintegrate, losing its insulation blankets in bits and chunks. I think the adhesive holding the blankets to the outer wall of the spacecraft was falling apart under the constant bombardment of solar ultraviolet rays. COBE, it appeared, was getting a severe sunburn. The spacecraft's electronics remained switched on as we handed its controls over to NASA communications engineers at Wallops Flight Facility, who planned to test their ground stations with signals from its transmitters. But no more scientific data would ever be sent back to Earth from its instruments.

COBE is now merely one more small chunk of the universe it had explored so successfully. It should remain in orbit for a thousand years or more. Eventually the minuscule air drag on the spacecraft will pull it down through the atmosphere like a fiery meteor. That could by then be the only remnant of our age's effort to comprehend the universe.

By the time COBE was closed down, we had four years of good

data in the Goddard computers; obtaining more would not have justi-fied the additional expense. In those four years the FIRAS team had greatly improved the calibration software through the efforts of Dale Fixsen and Dave Cottingham, giving us more confidence about the data. The team wrote four papers: one giving a summary, a second reporting on the cosmic significance of our spectrum, the third about measurements of the dipole, and a fourth about the incredibly difficult calibration software.[3] My colleagues and I presented the results at the January 1993 meeting of the American Astronomical Society in Tucson.

The FIRAS papers received little notice other than a front-page story (below the fold) in the New York Times that stated: "Astrophysicists said the infrared measurements provided further confirmation of the basic Big Bang theory as the best explanation of how the universe began some 15 billion years ago."[4] Perhaps reporters were beginning to think that the Big Bang was perfect, confirmed, and, hence, boring. British theorists Martin Rees and Malcolm Longair, however, believed that confirmation of the FIRAS spectrum was even more surprising than the DMR anisotropy: The evolving universe had managed somehow to produce galaxies and galactic clusters while releasing only a small level of far infrared radiation even though many theories pre-dicted that a large amount of such radiation should still be present.

It took until 1996 before the FIRAS analysis of the spectrum was completed. Dale Fixsen, who had led the calibration effort, showed that the cosmic spectrum matched the prediction within an incredible 0.01 percent. This was over ten times better than we had promised in the beginning. It means that no more than 0.006 percent of the energy in the background radiation was added after the first year of the the expanding universe.[5]

Still analyzing the wealth of DMR data after COBE was shut off, most of us on the science team believed the remaining uncertainties related not to the spacecraft's instruments but to the cosmos. The big problem was that the cosmic background radiation appeared to have a random pattern of temperature fluctuations spread across the universe. This means the COBE data could be deceptively limited, since the space-craft viewed the cosmic hot spots only from our single Earthly perspec-tive. Such an uncertainty is similar to the problem of determining whether a coin toss is fair from just a few flips. Randomness was inherent to the COBE's view of the universe, and its detectors could only measure the

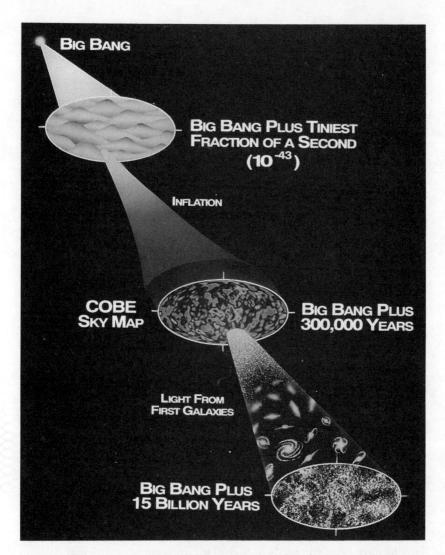

Fig. 13. ARTIST'S CONCEPT OF WHERE COBE FITS IN BIG BANG
COSMOLOGY.

The first oval represents the universe just after the Big Bang, when the universe was very small and very hot and contained irregularities in the structure of space. This was followed by a period of rapid inflation. The second oval depicts the COBE differential microwave radiometer sky map after one year of observations and represents the universe about 300,000 years after the Big Bang. This is when the universe had cooled down enough to become transparent to radiation and when galaxies and stars could begin to form. The third oval represents the universe of today, consisting of billions of galaxies.

largest thousand or so spots in the sky as seen from here. (See figure 13.)

The quandary existed at the start of the project and remained at the end. When my colleagues and I first conceived COBE, we were determined to reach the limits set by our location in the universe. It was impossible for us to send a spacecraft outside our own galaxy, or even out of the solar system, hence our constant worry about galactic noise. Our aim was to do the best we could, living in our little corner of the cosmos. By the time COBE was shut down, it had achieved that goal.

The COBE spacecraft has opened an era of intense enthusiasm among scientists for the study of the cosmic background radiation they believe is the afterglow of the creation of the universe. As COBE radioed its final reports back to Earth, at least a dozen new experiments were being planned to study the radiation from the Big Bang at even finer angular resolutions. Although most of these experiments were ground-based or would be sent aloft in balloons, NASA received three serious proposals for new satellite observations of the cosmic background radiation. One, entitled the MAP (Microwave Anisotropy Probe) and led by Chuck Bennett and Dave Wilkinson, has been selected. European scientists are planning an even more ambitious effort in space called the COBRAS/SAMBA. Russian scientists, too, were contemplating a proposal to launch a second *Relikt* satellite they hope will make a finer measurement of the cosmic afterglow.

Phil Lubin, a COBE alumnus, was conducting another experiment based in Antarctica, where the cold atmosphere provides the clearest view of the cosmos from Earth. If his detectors work as designed, they could find temperature fluctuations nearly twenty times smaller than those detected by COBE. These would be tiny enough, theorists calculate, to have been the seeds in the infant universe of individual galaxies such as our own Milky Way. I have been involved in proposals for two more space missions that would greatly refine COBE's other results. One experiment would make a longer wavelength spectrum measurement, while the other would attempt to improve upon DIRBE with an instrument aboard a mission beyond the interplanetary dust cloud.

Two decades ago, when my colleagues and I first conceived a mission in space to study the cosmic background radiation, there was none of the intense interest of today. Oh, the radiation was important enough. But it was difficult, if not impossible, to measure further.

Other astronomical phenomena found in the infrared or X-ray regions of the universe would provide us with far more clues about the nature of the cosmos, it was believed. Now, owing to the persistence of the COBE team and strong-willed individuals such as Nancy Boggess, Ray Weiss, Mike Hauser, Roger Mattson (who died along the way), Dennis McCarthy, and others, cosmologists realize that the cosmic background radiation provides us Earth dwellers with our clearest snapshot of the events that took place billions of years ago when the universe was very young.

COBE discovered that the cosmic background radiation has a virtually perfect blackbody spectrum. It found minute anisotropy in the cosmic background radiation, the mysterious missing link in the story of the universe's evolution. Did COBE "prove" the Big Bang, as many popular accounts have suggested? I don't have that answer any more than the authors of these articles and books. Chuck Bennett held up such an article one day. "You just can't prove the Big Bang," he said, laughing. Will we humans ever know absolutely how the universe began? Probably not. After all, an event such as the Big Bang, if it indeed occurred, took place in the vast unknown reaches of the most distant past, far beyond the direct vision of us humans and our finest instruments.

What do we know with certainty? Only that the universe exists and is vaster than we had ever imagined: a great firmament of galaxies and quasars, nebulae and dust, luminous stars, planets, and people, along with an unknown if not unknowable dark matter, intersecting gravitational fields, and powerful force fields deep within the fundamental particles of nature—all endowed with an evolving creative power of both great simplicity and great complexity, suffused with an ancient glow of magnificent uniformity and surprising beauty and containing, in our scientific concept of creation, the seeds of everything to come.

EPILOGUE

The sole purpose of human existence is to kindle a light in the darkness of mere being.

— CARL GUSTAV JUNG

THINK ABOUT ONE OF the photographs taken by an astronaut on the Moon. The Earth is a brilliant blue and white sphere suspended in the blackness of the night sky. Now back up to a more distant part of our solar system, somewhere out past Mars. Imagine what the earth would look like from there. It probably would resemble Venus or Mercury in the evening sky, a bright, unflickering star. From the black depths of the solar system near Neptune or Pluto, the Sun would dim, and our home planet would disappear into the night.

Journey out of the solar system to the nearest visible star, Alpha Centauri, 4.3 light-years away. Now the Sun is just another point of light in the sky. Keep traveling: out into the galaxy as the Sun disappears into a cloud of stars, then beyond the Milky Way, zapping at warp speed (it would indeed have to be an exotic form of travel) two million light-years to the nearest large spiral galaxy, Andromeda. Where is the Sun now? Could you locate our planet in the cosmic abyss?

While you make your imaginary cosmic voyage, ask yourself one question: Was all this vast universe created just so people like us could live, work, play, and wonder about things like this on Earth? Some cosmologists who have helped develop several versions of the so-called anthropic principle answer such a question in the affirmative. The principle, in its simplest form, asserts that current conditions in the universe—including the existence of planets such as Earth, the stars, the galaxies, and the huge galactic structures, along with the measured strengths of gravity and the three more powerful electromagnetic and subatomic forces—all have coalesced to allow for the existence of life. A stronger version states that *only* in such a universe could human life have emerged.[1]

The anthropic idea has seduced thinkers for a long time. Alfred Russel Wallace, a colleague of Charles Darwin's, suggested in *Man's Place in the Universe* that had the earth and solar system been constructed differently, conscious life would have been impossible. In a wonderfully irreverent essay called "The Damned Human Race," which appeared posthumously in *Letters from the Earth,* Mark Twain dispatched Wallace's idea: "If the Eiffel Tower were now representing the world's age, the skin of paint on the pinnacle-knob at its summit would represent man's share of that age; and anybody would perceive that that skin was what the tower was built for. I reckon they would. I dunno."

I don't know, either. Nor does anyone. Cosmologists have a good idea *how* the universe may have evolved, but not *why.* After all, "purpose" is not a measurable physical quantity like length or weight. In the version now current, the earliest phase of the universe was astonishingly hot and dense, so much so that the laws of physics were obscured and the traditional meanings of space and time were distorted by the curvature of the gravitational field. The idea of a "creation event" is now being generalized to a process that could happen over and over. The idea is plausible, I think, but entirely untestable in any way we can imagine now. As matter in the early universe cooled and expanded, unstable elementary particles decayed, releasing energy. After the temperature dropped to a few billions of degrees, some neutrons joined with protons, creating helium nuclei.

The hypothetical dark matter started to fall onto the primordial seeds of gravitational fields. After another 300,000 years the temperature dropped to about 3,000° Kelvin. As electrons joined atomic nuclei to make atoms, the universe became transparent and the new atoms became free to move within its gravitational field. Early clumps or proto-galaxies started to appear, and these may have begun producing stars from the primordial material. The earliest stars contained almost no atoms heavier than hydrogen or helium, while later stars included carbon, oxygen, nitrogen, and other heavier atoms generated in nuclear reactions. Today one of astronomy's real arts is determining the age of a star by studying the abundance of its elements and its structural details.

Our Sun, about 4.5 billion years old, appeared late in the history of the universe. It may have been enriched by a nearby supernova that blew up, releasing its nuclear products such as iron, copper, and all the other materials we use now for our technology. The Sun formed a few hundred thousand years later, with the earth coalescing about as quickly from supernova debris not drawn into the Sun. The early Earth was an awesomely strange place. The Sun was faint in the sky, and planetary debris rained down on the surface. The atmosphere was mostly carbon dioxide and nitrogen, with very little oxygen. The surface was a liquid soup of chemicals. Some combination of these chemicals must have begun to stabilize and grow larger.

According to the modern theory of complexity and chaos, such an event was likely to have occurred quickly and many times (natural processes such as the weather that use energy flows to build complexity are quite common). Once a structure capable of random variations and inheritance developed, its complexity could grow astonishingly fast under the influence of natural selection. Thus life began (maybe).

Oddly, there is little evidence of rapid change in the geological record of the earliest life forms. Evidence for fast-paced evolutionary change in later, more developed life is abundant, such change often assisted by catastrophic events. Some 100 large craters on the earth's surface testify to the crash of comets and asteroids. An impact that could have wiped out the dinosaurs and other life forms about 65 million years ago may have left such a crater on the Yucatán peninsula. In this century a vast forest around Tunguska in Siberia was devastated by the impact of a comet head of which no trace has been found. The evidence indicates that plants and animals evolved very rapidly after such drastic environmental changes. In modern times the pace of evolution is so extraordinary that in just a few thousand years of human— rather than natural—selection, every domestic animal and plant has been bred. Most bear little resemblance to their wild ancestors. We humans have existed in our modern guise for only a few hundred thousand years, a flash in the cosmological pan.

Did humans evolve as a widely distributed population? Or did we evolve in isolated groups? Theories exist, but nobody really knows.

What is known is that the earliest human prototypes had a few special characteristics they shared with modern people: color vision (like birds but not other mammals) and true stereoscopic vision, with the nerve from each eye dividing to both hemispheres of the visual cortex. With such traits and others, the earliest hominids separated from their chimp and gorilla cousins several million years ago. As their brain capacity increased their survival abilities improved, and they were able to move into more hostile regions.

Today not all human groups are the same. There are significant differences in the internal structures of our bodies, while biochemistry is known to vary from one race to another. Recently it has been found that Australian aborigines have as a group a detectably different brain function and structure from other humans. Despite such evidence, nobody has been able to show at what evolutionary stage these kinds of differences first appeared.

All of us arrived in the present time as the result of evolution's essential feature: its neutrality. Living in a natural world that did not care—indeed, could not care—who lived or who died, our forebears beat out their opponents in the deadly game of survival. Today this so-called survival of the fittest sometimes is used to justify the privilege and power of the human species on our planet: We won the war of evolution, so we can do what we want. This is a deeply flawed view. In the survival of one species over another, there is only one truth. "Winning isn't the most important thing. It's the only thing," said football coach Vince Lombardi. The remark as it relates to evolution is not a moral or ethical comment. It only states the obvious: A species that fails to produce will not have any descendants—a tautology.

"Let them have dominion over the fish of the sea, and over the fowl of the air, and over the cattle, and over all the earth," God said of humans in Genesis. It is easy for us to convince ourselves of the statement's truth today. Our species has an ability to change the planet unmatched by any other life form since free oxygen was released into the atmosphere. Only a few thousand years after the invention of agriculture, we have cultivated a large fraction of the earth's arable land. The wild forests are mostly gone. Species are becoming extinct daily as their habitats are consumed by other species—humans or other species such as English sparrows and starlings that can coexist well with us. The dominion of nearly five billion humans over everything else that lives appears total.

THE INVISIBLE HAND AND LONG-TERM SURVIVAL

Evolutionary victory has never been final. Mother Nature is not pretty, sweet, or orderly. Chaos, instability, and mystery rule. Our successful ancestors and their unsuccessful opponents suffered death and destruction. Over millions of years the planet's geology and climate have changed catastrophically; asteroids and comets have struck; and evolution has occurred as a means of accommodating these changes, not the other way around. It is estimated that there are now 50 million species of plants and animals on Earth. Yet it is believed that since life began, there have been 50 *billion* species. Hence for every thousand species that have existed, 999 are now extinct. The average lifetime of a species has been about four million years.

Although we may think otherwise, modern humans have no biological claim to a special status. We have not reckoned on the opposition that nature, in the form of a great catastrophe or an uncontrollable virus, or ourselves, could bring against us at any moment. Science holds no value judgment. It notes only that over geological—not to mention cosmic—time scales, all will be erased, burned up by the expanding Sun or eroded and covered with miles of sediment on the earth's surface.

Still, humans are unique. Of all nature's creatures (of which we're aware), only we have the power to contemplate our behavior and alter it to achieve a different end. Is the true meaning of the expulsion from Eden that humans could no longer claim not to have those powers? An interesting speculation is to consider the challenges facing a species wanting to achieve immortality here on Earth. It is clear that those of us living today cannot make decisions for future generations. What we can do is attempt to build the foundation for a life-sustaining civilization with the capability of running forever.

Are we doing that now? What would an eternal species require? Two things seem obvious. The first—a continuously improving technology for the production of material goods and services—we already have. The other is a long way off—an ability to solve large-scale problems affecting the entire species. The problem is simply stated. There is no guarantee that individuals and groups behaving as they must to survive will cooperate to achieve survival of the species as a whole. Today there is no mechanism for preventing individuals or groups acting in their own interests from using up or abusing all the collective resources, leaving nothing for future humans.

Does anybody have any idea how to achieve the long-term survival of humans? In the West we have put our faith in the "invisible hand" that has guided economic theory ever since Adam Smith. Recent trade agreements have been motivated by a faith among economists and the government officials who listen to them that more trade guided by an "invisible hand" is better. Yet voters in democratic nations often do not agree, fearing the increased competition—or survival pressure—open trade brings. Who is right? There is no proof of the validity of the invisible hand for the perpetual survival of the species.

Karl Marx abandoned the invisible hand for another kind of decision-making process relying heavily on centralized control. His followers in the Soviet Union and elsewhere claimed that they were testing his theory. Yet such control, instilled by many regulations, rules, and easily corrupted bureaucratic agencies, has led repeatedly to economic catastrophe because it is powerless to stop an ongoing onslaught of external change. In Mexico the government spent all its reserves attempting to maintain a fixed dollar-peso exchange rate. When the reserves were expended, the economy collapsed. Central controls have failed in Cuba and Eastern Europe. The Soviet Union collapsed after decades of a failed experiment in which governmental structure attempted to follow an ideal opposed to more powerful forces of human nature. Efforts to achieve long-term stability by means of centralized regulation and organization facing unknown future opposition may simply be a bad strategy that leads inevitably to a catastrophic outcome. Natural systems, human societies included, seem to find a weakness in any armor.

No one—Karl Marx, Adam Smith, or anybody else—has created a measurement of what is "best" for the permanent survival of our species. Obviously, a measurement based on future consequences will not work. We cannot predict even the weather for more than a few days, much less anything that will happen years, centuries, or millennia from now. Hence, current prescriptions for behavior, ethics, and morality are based on principles that relate not to species survival but to individual or group survival. Several organizing principles are used by most societies: law, established by force or agreement; religion, established by persuasion; and culture, established by custom and experience. Each is vitally important and, since the stakes are immense, generates constant pushing and shoving within a group or between groups. But each can be readily abused by the powerful or persuasive.

That leaves scientific inquiry, alone among human enterprise a neutral force. Barring a recurrence of the Dark Ages, the accumulation of scientific information is an irreversible process. A scientist's sacred duty is to add to this store of knowledge. I am fascinated by libraries where such learning is reposited. The Bookmobile started my scientific education. I have visited the ancient library at Ephesus, a three-story stone building that held the essence of the Western world's knowledge in biblical times. In Córdoba I saw where Arab scholars kept religious tolerance and scientific knowledge alive during a time when Europe was darkened by poverty and ignorance. I am equally fascinated with the Library of Congress, the New York Public Library, guarded by the famous lions, Patience and Fortitude, and Jefferson's library at Monticello.

Submitting a manuscript to the *Astrophysical Journal,* thereby adding a kernel of knowledge to the world's libraries, is, to me, an indescribable thrill. It comes from knowing that one is playing a part in the ongoing cosmic library drama. One may never be able to trace the ripples engendered by one small individual action among the many millions in the great human drama. But adding to a library comes as close to immortality (at least as far as our species is concerned) as the ordinary person can come.

Scientists invariably are asked to explain why their work is important when people around the world are starving. Of course, one cannot state that more people will eat well tomorrow because I have submitted a paper on the origin of the universe to the *Astrophysical Journal* today. Yet one reason that many people are starving is that individuals lack knowledge of a generally accepted means of producing and distributing food where it is needed. That knowledge, which already might exist somewhere in the collective human library, would be a great power if everyone could have it. Producing more knowledge may be the only long-term means of keeping people from starving. Who would have imagined that direct broadcast satellite television would lead to improved agriculture and health in India and the former Soviet Union?

Conversely, of course, a technology that enables all five billion humans to consume the planet's resources even faster may not prove to be a good long-term survival strategy for the human species.

QUO VADIS, SCIENTIA?

Governments have sponsored scientific research ever since (at least) the Tyrant of Syracuse employed Archimedes to develop military machines. In the twentieth century, government support, motivated by the two world wars and the cold war, has led to truly remarkable advances. In military competition, knowledge really is power, especially knowledge not shared by one's enemies. Now the commercial sector also invests heavily in scientific research, particularly in those areas that could have payoffs within a few years. With competition shifting away from research for purely military applications, it is appropriate to wonder what role government should take in supporting and directing science.

Clearly, no nation can successfully ask individuals or corporations to pursue or restrict scientific research for a greater societal good or for the long-term survival of the species. The question should be *Cui bono?*—Who benefits from the work? There always will be scientific questions simply too difficult or of limited commercial benefit. In a democracy working well, government represents the largest-scale business enterprise, manufacturing and selling products and services to the buying public (also known as taxpayers). If the public wants scientific research, the government should provide it. If the government does a bad job in providing such a service or any other, the government should be removed. Fortunately, there exists in the United States a formal competitive process for governmental funding of scientific projects that values originality and new perspectives. Had such a system not been in place, COBE would never have been launched.

In the mid-1990s NASA, under the same congressional committee as housing and veterans affairs, has been under attack, its budget cut by a third. With the exceptions of the planned space station and an Earth-observing system, only small projects were being considered. Yet only NASA can undertake certain large scientific projects, such as looking back at the beginning of the universe or speculating about the future as we monitor the loss of the ozone layer. NASA has opened new windows on the cosmos, with the Hubble and the Gamma Ray observatories, and on the earth, by monitoring the size of the polar ice caps to determine whether global warming is significant. If the earth's environment can no longer sustain us, we eventually may be able to travel back to the Moon and on to Mars and colonize them. First, though, far more powerful computers will be needed to create detailed simulations of such trips and to organize human management.

The COBE project has been funded through 1997 for analysis and data production, but its staff has been shrinking as the work is completed. Scientific attention on the cosmic background radiation has shifted to measuring even smaller-scale anisotropies. The 1° scale fluctuations have been measured with great confidence, although no measurement exists for much smaller scales.

Our theoretical colleagues who study the growth of cosmic structure from the primeval seeds found by COBE now claim they have figured out the mechanism of the early phase of expansion. Detailed simulations can now show that the initial conditions measured by the DMR really could eventually have produced both the large and small structures now seen in the universe. The question once was: Do small structures form first and then aggregate or do large structures form first and then divide? The new simulations based on our data show that both occur at the same time. For the first time, theory, computer simulation, and observation are all in agreement.

Theorists now realize that emerging details of the cosmic background radiation's anisotropy may be even more informative than they first realized. When ordinary matter started to move toward clumps of dark matter at the moment of decoupling, the cosmic background radiation would have been profoundly affected: The matter falling into the dark matter probably oscillated in and out several times. Further measurements of the cosmic background radiation could reveal details of these oscillations. This in turn should tell us about the density and properties of both ordinary matter and dark matter. Equally fascinating, we might be able to determine the total density of the universe. Our recognition of such fluctuations could be the most direct evidence we ever have about the nature of the Big Bang and the laws of physics operating under its extreme conditions. Theorists such as Stephen Hawking seeking the great Theory of Everything believe such measurements could lead to their Holy Grail. Four major studies are under way to seek suitable projects to make additional measurements. I am hopeful that at least one eventually will be done.

Traditional astronomy is also taking big new strides. Advanced technology from the military and commercial sectors has become available following the downfall of the Soviet Union; better ground-based astronomy can be done for less money. An article in the *Astrophysical Journal* entitled "Quo Vadis, Astronomia?" laid out the future of astronomy: Large telescopes will improve dramatically as they incorporate the image-stabilization technology developed for the Pentagon, then subsequently declassified.[2] This technology relies on flexible mirrors to compensate for distortions caused by atmospheric turbulence. The technology is so powerful that it will bring remarkable advances in our knowledge without the great cost of building new telescopes.

Still, even at a high mountain site the atmosphere is too opaque for the wavelengths of greatest interest to astronomy. For these wavelengths, from a few micrometers to a millimeter, space is the only place where we can acquire meaningful new data. NASA already has sent small telescopes into orbit; the next step is larger ones, which, of course, cost more. Unfortunately, NASA is no longer science's rich uncle. But astronomy should benefit from ongoing technological advances made elsewhere. Sending larger and larger detectors into orbit will mean relying on more advanced computers to keep the price down, the cost of human thought still the largest expense of getting into space. Fortunately for science, we are in the midst of a telecommunications and computer revolution. Computer-aided design and manufacturing should become advanced enough to enable far larger rockets and telescopes to be built for a reasonable price.

NASA has begun studying a successor to the Hubble Space Telescope. I have been appointed Study Scientist, to work with colleagues throughout NASA and in industry, universities, and the Space Telescope Science Institute, to develop a plan. We hope to use new technology to enable the Next Generation Space Telescope to reach much farther than ever before, at much lower cost.

SCIENCE VERSUS RELIGION?

The story of creation in Genesis places events in approximately the same order as the modern scientific version of the beginning of the universe. Other than that rough agreement, the old religious texts do not paint a picture of the universe matching that of modern science in other details. Of course, the writers, translators, and copiers of the old books did not have our concepts of general relativity and quantum mechanics to help them understand the universe.

There has been, I think, a long and unnecessary confusion over the difference between religion and science. Some of the confusion has been intentional, since proselytizers on each side use any argument at their disposal. Religion and science, however, really are about entirely different domains of human experience that only appear to share a common language. The conflicts that can arise at the boundary between religious and scientific views of the physical universe are demonstrated by the lingering controversies over the scientific work of Darwin and Galileo.

Today many educated individuals may not be aware that the Vatican has operated its own observatory for years, or that one of the founders of the Big Bang theory was a priest who was also a mathematician (or was it the other way around?), or that the Roman Catholic Church is often the sponsor of conferences on scientific cosmology. On May 8, 1993, Pope John Paul II visited the Ettore Maiorana Center in Sicily, a conference center for scientists in Erice.

"Faith and Science Both God's Gift," stated L'Osservatore Romano's headline of the report about the visit (May 9, 1993). "The Galileo case has been a sort of 'myth,' in which the image fabricated out of the events was quite far removed from reality," the Pope was quoted as saying. "A tragic mutual incomprehension has been interpreted as the reflection of a fundamental opposition between science and faith." Historical documents examined by scholars indicate that Galileo's conflict with the church had little to do with the Copernican revolution, as most of us think today because of Bertolt Brecht's play Galileo. In fact, the persecution of Galileo was deeply involved with events of the day, such as the challenge of the Protestant Reformation and the debate within the Church leadership over how to respond to it. Then as now, hard-liners and reformers battled. Galileo's case was caught in the middle of these larger forces, and he was convicted of heresy by the Inquisition in 1633. Three and a half centuries later, Pope John Paul II asked the Church to annul the conviction.

So, do scientists believe in God? The subject rarely seems to come up as a topic of conversation among scientists I know. But I occasionally visit churches, talking about COBE's discoveries. I invariably find a receptive audience, curious to puzzle out the meaning of life and the universe for themselves. My answer is always the same: The questions about God, science, and creation are themselves provocative, and the partial answers we have are fascinating, and the scientific discoveries are thrilling.

Will the advancing astronomical technology mean that some day we'll know with certainty whether today's exotic cosmological theories about the beginning of the universe are right? Probably not. As far as we observational astronomers can tell from our measurements, the origins of the early expanding phase of the universe were so extreme that they may have erased traces of earlier events, which in turn perhaps involved unknowable laws of physics. Our best hope may be pure thought in the form of mathematics bound tightly to physical observation and experimentation.

Perhaps one day we will have imagined the right theory, and we will think it so beautiful that it must be true. Of course, it is just as likely that the real truth is so complicated or obscure that we cannot discover it or recognize it even if we encounter it directly.

The meaning of life and the universe may be beyond mortal comprehension. As J. B. S. Haldane wrote in *Possible Worlds* (1927), "The universe is not only queerer than we suppose, but queerer than we *can* suppose."

APPENDIX

COBE Science Team

Charles L. Bennett, GSFC, Deputy Principal Investigator for Differential Microwave Radiometers, DMR

Nancy W. Boggess, retired from Deputy Project Scientist at GSFC and Program Scientist, NASA

Edward S. Cheng, MIT, then GSFC, was Deputy Project Scientist

Eli Dwek, GSFC

Samuel Gulkis, JPL

Michael G. Hauser, GSFC, Principal Investigator for Diffuse Infrared Background Experiment, DIRBE; now Deputy Director of Space Telescope Science Institute

Michael A. Janssen, JPL

Thomas Kelsall, GSFC, Deputy PI, DIRBE

David T. Leisawitz, GSFC, Deputy Project Scientist

John C. Mather, GSFC, Project Scientist and Principal Investigator for Far Infrared Absolute Spectrophotometer, FIRAS; now Study Scientist for Next Generation Space Telescope

Stephan S. Meyer, MIT, then University of Chicago

S. Harvey Mosely, Jr., GSFC

Thomas L. Murdock, Air Force Cambridge Research Laboratory, then General Science Corporation

Richard A. Shafer, GSFC, Deputy PI for FIRAS

Robert F. Silverberg, GSFC

George F. Smoot, Lawrence Berkeley Laboratory and University of California, Principal Investigator, DMR

Rainer Weiss, MIT, Chairman of Science Working Group

David T. Wilkinson, Princeton University

Edward L. Wright, MIT, then University of California, Los Angeles, Data Team Leader

COBE Engineering and Managing Team in 1988

Donald F. Crosby, Instrument Engineer, DIRBE

Ernest C. Doutrich, Flight Assurance Manager

Irene K. Ferber, Project Secretary

Anthony D. Fragomeni, Observatory Manager
Thomas J. Greenwell, Integration and Test Manager
David Gilman, Program Manager, NASA Headquarters
William D. Hoggard, Delta Liaison/Launch Operations Manager
Charles Katz, Systems Engineer, Instruments
Bernard. J. Klein, Instrument Engineer, DMR
Loren R. Linstrom, Systems Engineer, DIRBE
Robert J. Maichle, Instrument Engineer, FIRAS
Robert A, Mattson, Project Manager
Dennis K. McCarthy, Deputy Project Manager
Maureen J. Menton, Secretary
Herbert J. Mittleman, Resources Officer
Stephen Servin-Leete, Systems Engineer, DMR
Earnestine Smart, Project Support Specialist
Pierce L. Smith, Ground Data Processing Systems Manager
Jack W. Peddicord, Deputy Project Manager (Resources)
Michael Roberto, Systems Engineer, FIRAS
Robert G. Sanford, Mission Operations Manager
Robert T. Schools, Project Support Manager
Joseph F. Turtil, Systems Engineer
John L. Wolfgang, Software Systems Manager
Earle W. Young, Instruments Manager

Contributors to COBE

Eve Abrams
Rose Acevedo
Norm Ackerman
Mario Acuna
Charles Adams
Kenton Adams
Susan E. Adams
W. Adams
Mary Adkins
Darlene Ahalt
Eliezer Aharonovich
Angela Ahearn
Anisa Ahmad
Jim Akers
Cheryl Albert
Robert Aleman
Steve Alexander
Bruce Allen
Phillip Alley
Calvin Allison
Mary Aloupis
David Amason
Wendy Ames
Matthew Anderson
Tony Andoll
L. C. Andreozzi
Earl Angulo
William Anonsen
Mark Arend
Nerses M. Armani
Robert M. Armstrong
Petar Arsenovic
Lois Aylor
Jon Aymon

Karin Babst
Neal D. Bachtell
Charles Backus

Yoon Bae
Susan Bailey
Audrey J. Baker
Don Baker
Barry Baltozer
Melvin Banks
Dale F. Bankus
Quinton Barker
Joe Barksdale
Bill Barnes
Jeff Barnes
Rich Barney
Andy Barr
Harold S. Barsch
D. E. Bartels
Michael Barthelmy
Shivanand Basappa
Lisa Basiley
Mark Baugh
Robert Baumann
Phyllis Bayne
Dave Bazell
Earl Beard
Russ Beard
John M. Beckham
Howard Becraft
Tony Bello
Porfirio Beltran
Jeannette Benavides
M. D. Bengtson
Chuck Bennett
Jerome Bennett
Joe Bentley
Darlene Bently
Ronald L. Bento
Graham Berriman
Colman W. Beulah
Narin Bewtra

John Bichell
Arlene R. Bigel
Sandra Biggs
Diane Blair
Wilfredo Blanco
Sharon Bland
Tom Blaser
Nate Block
Garcia Blount
Diane Bobak
Ed Boggess
Nancy W. Boggess
Jerry Bonnell
Carol Boquist
P. L. Borde
Francesco Bordi
Richard Bost
Ray Boucarut
Dave Bouler
Bob Bounds
Don Bower
Gordon Bowers
Gregory Bowers
William T. Bowers
Priscilla Bowes
Jeffrey Bowser
Rob Boyle
Barbara Bradtmueller
Regis Brekosky
Keith Brenza
Polly Bresnahan
Carrie Brezinski
Vincent J. Briani
Mildred Brice
Don Briel
T. C. Briggs
Lawrence Bromery
Marvin Brooks
Cecilia Brown
Heather Brown
Kim Brown
Leonard Brown

Mark A. Brown
Nate Bruce
James Buckeridge
Spike Bukowski
James Bullock
Chris Bunyea
Shawn Burdick
Mike Burgess
Carol Burke
Lynette Burley
Peter Burr
Valorie Burr
Doug Burritt
Helen Burritt
Joe Burt
Scott Burtis
Gretchen Burtton
Frank Bush
Tom Butash
James Byrd

Wally Cacho
Robert W. Cade
Jim Caldwell
Jody Caldwell
William Campbell
Malcolm Cannon
Darlene Capone
Hawk Carnahan
Armen Caroglanian
Dan Carrigan
Don Carson
Karen Carter
Denise Cartwright
Enrico Caruso
Paul Caruso
Chris A. Cascia
Bill Case
Michael K. Cassidy
Steve Castles
Jose R. Cerrato
Robert Chalmers

William Chambers
Carolyn Chandler
Po Lam Chang
Sysan Chang
Chim Chao
Hwai-Soon Chen
Ed Cheng
Wendy Cheng
Sarada Chintala
John Chitwood
Sang Choi
Jerry Christian
Kyle G. Christoff
Paul T. Christoff, Jr.
Se Chung
Quoc Chung
Larry Clairmont
David Clark
John Clarke
Carroll Clatterbuck
Steve Clementson
Bob Coladonato
Alex Coleman
Melanie Coleman
Dave Collier
Joe Colony
Michael W. Connor
Richard A. Conte
Lawrence Cook, Jr.
Jan Cooksey
Carolyn Cooley
Julie Corry
Ken Cory
Jack Coulson
Steve Cox
George Croft
Don Crosby
Leigh Croteau
Pat Cudmore
Bob Cummings

William Daffer

Lisa Dallas
Christopher Daly
Richard Dame
Charles Dan
Tony Danks
Nick d'Apice
Donald D. Davis
L. D. Davis
Mitchell Davis
Giovanni De Amici
Robert DeFazio
Don Deibler
Casey Dekramer
Tom Delaney
Ronald C. Delbrook
Robert Denhardt
L. W. Derouin
Becky Derro
Ashok Desai
Ed Devine
Jennell Dewitt
John DiBartolo
Debbie Dionisio
Mike Dipirro
Leon Donde
Martin J. Donohoe
Dewey Dove
Maurice Dowdy
Bill Doyden
Ted Drabczyk
Donna Drummond
Joseph Ducosin
Cedric Duffield
Tom Dugan
Betty Dumas
Anthony Dunston
Jeff Durachta
Dale Durbin
Eli Dwek
Jim Dye
Edward Dyer

G. E. Eastman
Ray R. Edwards
Bill Eichhorn
R. Einertson
Mathew Elliott
Daniel Ely
Joe Emmons
Rogelio Emralino
Robert D. Endres
Charlie England
Gene Eplee
M. E. Erickson
Tom Erickson
Ra Ra Errera
Bill Evans
Randee Exler

Don Fadler
Roger E. Farley
Frank Fash
Matt Fatig
Mike Fatig
Francis Federline
Richard Fedorchak
Paul Feinberg
Michael Femino
Irene Ferber
Cas Ferenc
Nelson J. Ferragut
Paula Ferring
Donald Field
Orlando Figueroa
Debbie Filson
Kelvin Finneyfrock
M. Fitzmaurice
Dale Fixsen
W. H. Flaherty
Mark Flanegan
Lisa Flemming
Cathy Fleshman
Yury Flom
Jose Florez

Manuel Florez
Robert Flynn
Everett Fogle
Walter Folz
David Ford
Mary Ford
B. G. Fox
Tony Fragomeni
Marcia Frances
R. E. Francis
Gren Frazier
Richard Freburger
Marty Frederick
Immanuel Freedman
Richard Freeman
Henry Freudenreich
Patricia R. Friedberg
Edwin H. Fung

Ron Gagne
Joel Gales
Gur Gallent
John Galloway
Kevin Galuk
Frederick W. Gams
Ken Gardner
Raul Garza-Robles
James Gary
Pat Gary
Jim Gatlin
Mike Gauss
Joseph Gauvreau
Gary Gavigan
Jay Geagan
Wayne Geer
Tim Gehringer
Guy Germana
Earl C. Gernert
Judy Gibbon
Carl Gieger
David Gilman
Carey Gire

Richard H. Gladfelter
E. L. Glenn
Kevin R. Glenn
Elva Glover
Gene Gochar
Meng Goh
James A. Golden
Carlos Gomex-Rosa
Nilo Gonzales
Walter Goodale
Cassandra Nix Goodall
Kris Górski
Bill Gotthardt
Ira Graffman
Randy Graham
Christie Grant
David Gray
Richard Gray
Charles Green
Walter L. Green
Linda Greenslade
Ron Greer
Wayne Greer
Jerry Greyerbiehl
George Griffin
Robert Grigsby
Frederick C. Gross
Michael R. Gross
Thomas Grubb
Kurt Grubler
Sam Gulkis
Jeff Gum
Ved Gupta
Paul Guy

John J. Habert, Jr.
Carl Haehner
K. K. Hagan
Sheila Haghighat
John Hagopian
Mike Hagopian
R. P. Haight

Thomas P. Hait
Milton Halem
Gardeania Hamilton
Paul Haney
Hagop Hannanian
Rich Hanold
William Happel
Thomas Harbach
Dick Harner
Abby Harper
Cheryl Harrington
Benjamin Harris
Gary Harris
Russ Harrison
Jon Harzer
Mike Hauser
Jim Heaney
Jeff Hein
Brian N. Helland
Hamid Hemmati
Gerald Hempfling
John Henegar
Paul Henley
John Henninger
David Hepler
Roberto P. Hernandez
Leon Herreid
Don Hershfeld
Tom Heslin
Patricia Hettenhouser
Paul D. Hill
Tom Hill
Larry Hilliard
R. Kenneth Hinkle
Noel Hinners
George Hinshelwood
Richard Hockensmith
Janet V. Hodges
Dick Hoffman
Henry Hoffman
Bill Hoggard
Howard Holbrook

Neil Holby
George Holland
Sue Holland
Wayne Holland
Richard Hollenhorst
Cecelia Hollis
Mike Hollis
Dave Hon
Michael A. Honaker
John Hopkins
R. A. Hopkins
Lynn Hoppell
Mike Horn
Elmer T. Howell
Qihui Huand
Tom Huber
David Huff
Allen D. Huffman
Peter Hui
Ed Hulbert
Carolyn Humphrey
Pat Humphrey
Ronald Hunkeler
Cle Hunt
Darlene Hunt
Jamil Hussain
Mark Hymowitz

Nick Iascone
Mary Igal
Jeffrey L. Ingream
Sandra Irish
Richard Isaacman
Harold Isenberg
Murry Itkin
Pat Izzo

Bill Jackson
Bob Jackson
Mattie Jackson
Pat Jackson
Pete Jackson

Dan Jacobs
Atul Jai
Joseph James
Mike Janssen
Bill Jarrell
Elizabeth M. Jay
Jane Jellison
Diane Jenkins
Eric Jenkins
Linda Jenkins
Kirk Jennings
Ken Jensen
Joel Jermakian
Jay Jett
Bernie Johnson
Clarence Johnson
Helen Johnson
Marva Johnson
Mike Johnson
Tammie Johnson
William Jolly, Jr.
Frank Jones
Isaiah Jones
M. E. Jones
Rick Jones
Shirley Jones
Willie Jones
Shawpin Jong

Ford Kalil
Sue Kaltenbaugh
George Kambouris
Bernard F. Karmilowicz
Gabriel Karpati
Sasha Kashlinsky
Chuck Katz
Lonny Kauder
Marvin Kaufman
Joe Keating
Phillip Keegstra
Michael E. Kefauver
Daniel Keith

Amy Kekeisen
Carl W. Kellenbenz
Linda Kelly
William Kelly
Tom Kelsall
Myron Kemerer
Steven Kempler
Cherry A. Kenney
Richard Kennon
Peter Kenny
James Kerley
Deborah D. Kershner
Bob Kichak
Bob Kidwell
Sang Kim
Charles W. King
Jamie King
Laurie Kipple
Frank Kirchman
Don Kirkpatrick
Emil R. Kirwan
Francis J. Kisner
Jom Kistler
Bernie Klein
Dan Klinglesmith
Douglas E. Knight
Lyle Knight
Dawn Knode
John M. Koenig
Alan Kogut
Diane Kolos
Andy Korb
Carl Kotecki
Bob Kozon
Betty Kramer
Bill Krause
Ronald Krellen
Mike Kridler
Jim Krise
George Kronmiller
Donald Krueger
Tim Krueger

Pete Kryszak
Mike Kulpak
Carol Kulwich
Vijaya Kumar
Bob Kummerer
Mike Kurtz
Louis Kurzmiller
Jonathon Kutler
Huang Kwei

Al Lacks
Carol J. Lacombe
Jim Lacombe
Annette LaJoie
John Lallande
Paul Lang
Rhea Langlois
Douglass J. Lankenau
Jim Lanks
Warren C. Lathe, Jr.
Connie Lau
Robert E. Laughlin
Kenneth W. Laverly
C. Laverty
Frank Lawrence
Jack Lawrence
Robert Leavel
Robert S. Lebair, Jr.
J. Lecha
Maria Lecha
Grace Lee
Norm Lee
Oleina Lee-Simpson
Henning Leidecker
Joel Leifer
David T. Leisawitz
D. D. Lemon
Michael Lenz
J. M. Lester
Brian Lev
Allen J. Levine
Doug Leviton

Ed Lewis
Willie Lewis
Robert Light
Michael Lin
Jerome Lindsey
John Lindsey
Charles Lineweaver
Loren Linstrom
Moussa Lishaa
Casey Lisse
D. E. Lloyd
Patrick K. Lo
Charles David Lockhart
Barry Lohr
Kathrine Long
Lois Long
Blake Lorenz
John Lorenz
Crockett Lowe
Barbara Lowery
Paul Loyd
Phillip Lubin
Ray Lundquist
Lorraine Lust

Snehavadan Macwan
Sean Madine
Larry Madison
Tim Madison
N. E. Magette
Tom Magner
Bob Maichle
Angie Majstorovic
Andrew Makar
William A. Mamakos
George Mangum
Kevin Mangum
John Mangus
Mark Mann
R. R. Manning
Dominic Manzer
Daniel Marinelli

Paul Marionni
Jane Marquart
Thomas H. Marsh
Dale Marshall
Tony Marshall
Bob Martin
Earl Martin
Frank Martin
Pilar Martin
Gregory Martins
John Maruschak
John Marvin
C. L. Maschauer
Shirley Masiee
James Mason
Derck Massa
John Mather
Eric Mathis
Andrew Mattie
Carmen Mattiello
Kelly Mattis
Roger Mattson
Houston Mayhall
Riley Mayhall
Mary S. Maxwell
Frank Mayo
Karen Mayo
Dennis McCarthy
Patrick McCaslin
Al McClannahan
Sid McClure
Matthew McCoy
Bill McDonald
Buck McDonald
Debbie McDonald
Gary A. McDonald
Mickey McDonald
Olivia McFarlane
Bill McGunigal
Dan McHugh
Chris McLeod
John McMahon

Joyce McMurtrey
V. B. McNeill
George McVeigh
John McWilliams
John Mejane
Nick Mejia
Fred C. Menage, Jr.
John Mengel
Maureen Menton
Darrell Merrill
Stephan Meyer
William H. Meyer
Joe Miko
Bruce Milam
Laura Milam
Andy Miller
Barry Miller
Bill Miller
Grace Miller
Henry Miller
Mark Miller
Paul Miller
Tony Miller
Jim Milligan
Rick Mills
Tom Mills
Linda Miner
James Ming
Peter Minott
William Mish
Darryl Mitchell
Joel Mitchell
Ken Mitchell
Linda L. Mitchell
Reg Mitchell
Mila Mitra
Herb Mittleman
Bill Mocarsky
Garu Moffatt
David Mok
Clarence R. Monn, Jr.
John Mooring

Armasndo Morell
Carolyn Morris
S. Harvey Moseley, Jr.
Carol Mosier
D. H. Mount
Charles Moutoux
Faye Movahhed
Paul Mowatt
Doug Mowen
Ron Moxley
Joe Muller
Ron Muller
James Mullins
Michael Mumma
James Munford
Alan Murdoch
Thomas Murdock
Ken Murphy
Paul Murray
William Myers

David Nace
Roy Nakatsuka
Cookie Namkung
Dick Nash
Elaine Nelson
Norman Ness
Doug Newlon
Son Ngo
Tu Nguyen
Larry Nichols
Harry Nikirk
Dave Niver
Leris M. Norman
Linda Norsworthy
Marci Norton
Prasad Nune

Nils Odegard
Ted Ogelsby
John H. Olesen
Leonard Olson

Terry O'Neill
Matt Opeka
R. M. Ortega
Edmond Oser
West Over
Linda Owens-
  Ryschkewitsch

Larry Pack
Frank Paczkowski
Stephen J. Paddack
Nancy Painter
David Palace
Aliza Panitz
Brad Parker
Brenda Parkinson
Jeta C. Parraway
Ziba Parsa
John Parsons
Bobby Patschke
Fred Patt
George J. Pattison
Nancy Patton
Steve Patton
John Paulkovich
D. A. Payne
Jack Peddicord
Jeffery Pedelty
David Perretin
James C. Perry
Bob Peterson
Dave Pfenning
Dung Pham
Minh C. Phan
James E. Phenix
Bob Phillips
John Phillips
George Pieper
Brian Pierman
Ryszard L. Pisarski
Samuel Placanica
Walt Plesniak

John Pocius
James Poland
Darrel J. Poloway
Karen Pope
Brad Poston
Angela Powell
Bernard Powell
Carl Powell
Gwendolyn Powell
Charles Powers
Edward I. Powers
Mike Powers
Urmila Prasad
Quain S. Prather
Larry Pratt
Henry Price
Robert Price
Dave Puckett
John Pyle

Edward R. Quinn, Jr.

Marie Rabyor
Atul Rae
Paul Ragan
Rebecca Ragusa
H. K. Ramapriyan
Rajendra G. Ramlagan
Ramjit Ramrattan
Abraham Ramsey
Roger Ratliff
Tom Ratliff
Lowell Rau
Anne Raugh
Shirley Read
Ron Rector
Yvonne Denise Reed
William J. Regan
J. K. Reidy
Leshun Relph
Al Renner
Bruce W. Rentch

Kirk Rhee
Cathy Richards
Carl Riffe
Carl Riley
Nancy Rinker
Katharyn Rivas
Michael Roberto
Dwight Roberts
George Robinson
James Robinson
Jon Robinson
Jill Rock
Eddie Rodriguez
Ernie Rodriguez
Otilia I. Rodriguez-Alvarez
Tom Roegner
James S. Rogers
Lonnie Joe Rogers
Laurie Rokke
K. G. Roller
Rich Rolnicki
Juan Roman-
   Velazquez
Francis A. Rondeau
Giulio Rosanova
Larry Rosen
Jacob Rosenberg
Bob Rosenberry
Peter Rossoni
Tony Rota
Shelley Rowton
Mary Royce
Jan Ruff
Art Ruitberg
George Rumney
Shirley Rupp
Tom Russell
P. Russo
Timothy Rykowski
Michael Ryschkewitsch

Debbie Sabatino

Willy Sadler
S. R. K. Vidya Sagar
Henry Sampler
Robert Sanchez
Bob Sanford
Edward Sanford
Stephanie M. Sanidas
Jairo Santana
Jay Santry
Carol Sappington
Mark Saulino
Clell Scearce
John J. Schaus
John Scheifele
Richard Schmadebeck
Gregory Schmidt
Sherry Schmitz
Bob Schools
Mike Schools
Evelyn Schronce
Diane Schuster
David Schwartz
D. P. Schwartz
John Scialdone
Courtney Scott
Jill Scott
Rick Scott
S. J. Scott
Kay Scoville
Diane Sebok
Lou Segar
Ritchie Seigel
Roman I. Semkiw
Nate Serafino
Art Serlemitsos
Debra Servin-Leete
Steve Servin-Leete
Sharon Shackleford
Rick Shafer
K. C. Shah
Charlie Shai
Rakeev K. Sharma

Oren Sheinman
Charles Sheppard
Allan Sherman
T. R. Shevlin
Bill Shields
Fred Shuman
Chris Silva
Allan Silver
Robert Silverberg
Joyce Simcox
Robert Simenauer
Carl G. Simon
Stephen Simonds
Alda Simpson
Doug Sims
Michael Singer
Kawal P. Singh
Ramesh Sinha
John Arthur Skard
Joseph T. Skladany
Rosalie Skrabak
Margie Small
Earnestine Smart
Jan Smid
Charlene Smith
James Smith
Larry Smith
Lee Smith
William Smith
William J. Smith
George Smoot
Edward Smygocki
Carl Solomon
Bill Spiesman
Chris M. Spinolo
Russ Springham
Walt Squillari
Bill Stabnow
C. H. Stahl
Rick Staveley
Joseph Stecher
Peggy Stevens

Don Stevens-Rayburn
Alphonso Stewart
John Stewart
Ken Stewart
George Stitt
N. D. Stoffer
Jack Stoner
Kathryn Stoner-Jhabvala
Al Strojny
Priscilla Struthers
Mary Jean Studer
John Sudey
B. A. Sullivan
Dave Sullivan
Edward Sullivan
Tom Sullivan
Steven G. Sutherlin
Betsy Sutton
John Sutton
Robert Sutton
Max Swantko
John Sween
Bob Swope
Joe Szczech
Wiktor Szczyrba

Darnell Tabb
Devin Tailor
J. J. Tarpley
Cynthia Tart
Mike Tasevoli
Gerald Taylor
Muriel Taylor
Krishna Tewari
Charles Thomas
John Thomson
Dick Tighe
Sid Tiller
Gary Toller
John Tominovich
Ray Topolski
Marco Toral

Hank Torrance
Eduardo Torres
Sergio Torres
Eduardo Torres-
  Martinez
Gregory Toth
Jack Townsend
Harry Toy
Alice Trenholme
Robert Trescott
Jack Triolo
Wen-Hu Tsai
J. L. Tucker
Joe Turtil
Tim Tyle
Nigel Tzeng

Joanne Uber
Bernard Uphsur
A. R. Urbach

D. A. Van Gundy
Tim Van Sant
Mike Van Steenberg
Jack Van Zant
Barbara Vargo
Frack Varosi
Edwin Vaughan
Ravi Venkataraman
Ralph Vidnovic
Michael Viens
Mike Vinton
Eugene Volpe
Steve Volz
R. G. Voorhees
Jan Vrtilek

Clarence Wade
James Wall
Nancy Walpole
Mark Walter
Ken Walters

Harte Wang
Doug Ward
John Ward
Shari Warner
Tim Warner
David K. Wasson
Sharon Watson
Robert C. Weaver
John Webb
Dick Weber
Janet Weiland
Ronald C. Weimer
Howard Weiss
Rainer Weiss
Dott Wells
J. A. Wells
R. F. Wendler
Don West
Steve West
David Wheeler
Benjamin White
Debra White
Richard White
Rickey L. White
Ray Whitley
Sandra Wigton
David Wilkinson
Bernadette Williams
Billy B. Williams
Domenica Williams
Stephanie Williams
Denise Wilson
Ed Wilson
John Wilson
Lisa Wilson
Reid Wilson
Julia Winkler
Howard Witcher
Chris Witebsky
Berl Wittig
Carl Wittman
Robert Wolde

John Wolfe
John Wolfgang
Gary Wolford
Jack Wolsh
Roger Wood
Steve Wood
Jim Woods
Natalie Woodson
William Woodyear
Lois Workman
Everett Worley
Lou Worrel
Barbara Wrathall
Ned Wright
Robert Wright
J. Andrzej Wrotniak
Don Wrublik
David Wyckoff
Dave Wynne

John Yagelowich
Jack Yambor
Jaya Yodh
Song Yom
Suk J. Yoon
Anthony Young
Earle Young
Pete Young
Stephania B. Young
Wesley Young

John L. Zahniser
John Zaniewski
Jane Zellison
Gerald R. Zgonc
David J. Zillig
Jerry Zimmerman

NOTES

CHAPTER 2: SCOOPED

1. B. M. Lynch and L. H. Robbins, "Namoratung'a: The First Archaeoastronomical Evidence in Sub-Saharan Africa," *Science* (1978): 766–68. A people called the Borano who have an unusual interest in timekeeping live across the border from Namoratung'a in the lowlands of southern Ethiopia today, and may have descended from the builders of the ancient Kenyan observatory; see C. L. N. Ruggles, "The Borano Calendar: Some Observations," *Archaeostronomy, Supplement to the Journal for the History of Astronomy* 11 (1987): 35.

2. Ludovico Geymonat, *Galileo Galilei: A Biography and Inquiry into His Philosophy of Science,* trans. Stillman Drake (McGraw-Hill, 1965), pp. 26, 31.

3. Telephone interview with Robert W. Wilson at the Harvard-Smithsonian Center for Astrophysics, Cambridge, Massachusetts, November 29, 1994.

4. R. W. Degrasse, D. C. Hogg, E. A. Ohm, and H. E. D. Scovil, *Proceedings of the National Electronics Conference* 15 (1959): 370, details the noise properties of the horn antenna, which had been studied in detail prior to Penzias and Wilson.

5. A description of this instrument is in R. H. Dicke, "A Microwave Radiometer for Measuring Cosmic Emissions," *Review of Scientific Instruments* 17 (1946): 268.

6. Interview with David T. Wilkinson, Princeton, New Jersey, December 20, 1994.

7. R. H. Dicke, R. Beringer, R. L. Kyhl, and A. B. Vane, "Atmospheric Absorption Measurements with a Microwave Radiometer," *Physical Review* 70 (1946): 340.

CHAPTER 3: THE REMAINS OF THE FIRST DAY

1. See Charles Coulston Gillispie's *The Edge of Objectivity: An Essay in the History of Scientific Ideas* (Princeton University Press, 1960) for a lively discussion of the thematic relationship among Tycho Brahe, Kepler, and Newton.

2. The most vivid account of Kepler and his work remains *The Sleepwalkers* (Columbia University Press, 1959) by Arthur Koestler, who observed, "There is hardly a page in Kepler's writings, some twenty solid volumes in folio, that is not alive and kicking."

3. Interview with Professor Rupert Hall, Oxfordshire, England, May 12, 1989.

4. J. D. North's *Isaac Newton* (Oxford University Press, 1967) contains a good account of the mathematics of the *Principia* for the general reader.

5. A. Einstein, "The General Theory of Relativity," *Sitzungsberichte der Preussichen Akad.d. Wissenshaften* 778 (1915); see also "Cosmological Considerations on the General Theory of Relativity" in the same journal, 142 (1917), both translated in Jeremy Bernstein and Jeremy Feinberg, eds., *Cosmological Constants: Papers in Modern Cosmology* (Columbia University Press, New York, 1986).

6. M. A. Markov, V. A. Berezin, and V. F. Mukhanov, *A. A. Friedmann: A Centenary Volume* (World Scientific, 1988).

7. A. Friedmann, "On the Curvature of Space," *Zeitschrift für Physik* 10 (1922): 277–86; A. Einstein, "Comments on the Work of A. Friedmann," *Zeitschrift für Physik* 11 (1922): 326 and 16: 228; and Friedmann, "On the Possibility of a World with Constant Negative Curvature," *Zeitschrift für Physik* 21 (1924): 326–32, both translated in Bernstein and Feinberg (see Bibliography).

8. Robert Smith, "Edwin P. Hubble and the Transformation of Cosmology," *Physics Today,* April 1990, 52–58.

9. James A. Bennett, "New Instruments Expand Man's Domain," in *Human Implications of Scientific Advance,* ed. E. G. Forbes (Edinburgh University Press, 1978), p. 553.

10. Smith, "Edwin P. Hubble," p. 53.

11. V. M. Slipher, *Lowell Observatory Bulletin* 58 (1913).

12. Smith, "Edwin P. Hubble."

13. Edwin Hubble, "A Relation Between Distance and Radial Velocity Among Extra-galactic Nebulae," *Proceedings of the National Academy of Science* 15 (1929): 168–73.

CHAPTER 4: THE DAY THE UNIVERSE CHANGED

1. G. Lemaître, "A Homogeneous Universe of Constant Mass and Increasing Radius Accounting for the Radial Velocity of Extra-galactic Nebulae," *Annals of the Scientific Society of Brussels* 47A (1927): 49, translated in *Monthly Notice of the Royal Astronomical Society* 91 (1931): 483–90.

2. Andre Deprit, "Monsignor Georges Lemaître," in *The Big Bang and Georges Lemaître,* ed. A. Berger (D. Reidel Publishing Co., 1984), p. 370.

3. G. Lemaître, *Nature* (suppl.) 128 (1931): 704.

4. John North, *The Measure of the Universe* (Clarendon, 1965), p. 126.

5. R. Alpher and R. C. Herman, "On the Relative Abundance of the Elements," *Physical Review D* 74 (1948): 1737.

6. R. A. Alpher, "A Neutron-Capture Theory of the Formation and Relative Abundance of the Elements," *Physical Review* 74 (1948): 1577.

7. R. Alpher, R. C. Herman, and G. A. Gamow, "Thermonuclear Reactions in the Expanding Universe," *Physical Review D* 74 (1948): 1198.

8. R. A. Alpher and R. C. Herman, "Evolution of the Universe," *Nature* 162 (1948): 774.

9. R. A. Alpher and R. Herman, "Reflections on Early Work on 'Big Bang' Cosmology," *Physics Today,* August 1988, 24–34.

10. H. Bondi and T. Gold, "The Steady State of the Expanding Universe," *Monthly Notice of the Royal Astronomical Society* 108 (1948): 252.

11. Interview with P. James E. Peebles, Princeton, New Jersey, December 20, 1994.

12. Telephone interview with Arno A. Penzias, December 1, 1994.

13. A. A. Penzias and R. W. Wilson, "A Measurement of Excess Antenna Temperature at 4080 Mc/s," *Astrophysical Journal* 142 (1965): 419–21; and R. H. Dicke, P. J. E. Peebles, P. G. Roll, and D. T. Wilkinson, "Cosmic Black-Body Radiation," *Astrophysical Journal* 142 (1965): 414–19.

CHAPTER 5: THE BALLOONING UNIVERSE

1. D. H. Martin and E. Puplett, "Polarised Interferometric Spectrometry for the Millimetre and Submillimetre Spectrum," *Infrared Physics* 10 (1970): 105.

2. John Cromwell Mather, "Far Infrared Spectrometry of the Cosmic Background Radiation," Ph.D. diss., University of California, January 1974; available from Xerox University Microfilms.

3. E. I. Robson, D. G. Vickers, J. S. Huizinga, J. E. Beckman, and P. E. Clegg, "Spectrum of the Cosmic Background Radiation Between 3 mm and 800 μm," *Nature* 251 (1974): 591.

4. R. Weiss and D. Muehlner, "Balloon Measurements of the Far Infrared Background Radiation," *Physical Review* D7 (1973): 326; and "Further Measurements of the Submillimeter Background at Balloon Altitude," *Physical Review Letters* 30 (1973): 757.

5. Telephone interview with Robert W. Wilson, November 29, 1994.

6. "Bell Scientists Detect a Cosmic Glow," *New York Times,* May 21, 1965, p. 1.

7. D. T. Wilkinson and P. J. E. Peebles, "How Not to Make a Discovery: A Case History," in *Serendipitous Discoveries in Radio Astronomy,* ed. K. Kellerman and B. Sheets, (National Radio Astronomy Observatory, 1983).

8. G. Gamow, "The Evolution of the Universe," *Nature* 162 (1948): 680.

9. R. Alpher and R. Herman, *Physics Today,* August 1988.

10. R. Alpher, G. Gamow, and R. Herman, *Proceedings of the National Academy of Sciences* 58 (1967): 2179.

11. Telephone interview with Arno A. Penzias, December 1, 1994.

12. Transcript of interview by Martin Harwit with Ralph A. Alpher and Robert Herman, archival material of the Niels Bohr Library, American Institute of Physics, College Park, Maryland, p. 78.

13. Ibid., p. 77.

14. Ibid., pp. 77–78.

15. P. J. E. Peebles and D. T. Wilkinson, "The Primeval Fireball," *Scientific American* (June 1967): 28–37. The article stated: "It is remarkable that this theory, as developed by Gamow, Ralph Alpher, Robert Herman and others, implied that the present temperature of the fireball would be about equal to the observed value of three degrees K" (p. 36).

16. Interview with P. James E. Peebles, Princeton, New Jersey, December 20, 1994.

CHAPTER 6: TUOLUMNE NIGHTS

1. William Broad and Nicholas Wade, *Betrayers of the Truth: Fraud and Deceit in the Halls of Science* (Simon & Schuster, 1991), p. 196.

2. Telephone interview with Arno A. Penzias, December 1, 1994.

3. A. G. Doroshkevich and I. D. Novikov, *Soviet Phys-Doklady* 9 (1964): 111.

4. A. McKellar, *Publication of the Dominion Astrophysical Observatory* (British Columbia) 7 (1941): 251.

5. Martin Harwit, *Cosmic Discovery: The Search, Scope, and Heritage of Astronomy* (Basic Books, 1981).

6. National Aeronautics and Space Administration, "Announcement of Opportunity in the Scientific Definition of Space Flight Investigations for Explorer Satellite Missions," Washington, D.C., July 15, 1974.

7. K. E. Tsiolkovsky, *The Investigation of Outer Space by Means of Reaction Apparatus* (1903), translated.

8. Esther C. Goddard, ed., *The Collected Papers of Robert H. Goddard,* 3 vols. (New York: Dutton, 1970).

CHAPTER 7: THE LAW OF INCIDENTAL CONSEQUENCES

1. A good history of NASA is T. Crouch's *The National Aeronautics and Space Administration* (Chelsea House, 1989).

2. See H. M. David, *Wernher von Braun: A Biography* (Meridian Books, 1967).

3. Interview with Rainer Weiss, Cambridge, Massachusetts, August 1, 1994.

4. Information relating to Soviet space successes and failures during the 1960s and 1970s comes from research and interviews conducted by John Boslough at Soviet space facilities in 1987 while on assignment for an article for *National Geographic* magazine, "Searching for the Secrets of Gravity," May 1989, 563–83. See also Thomas Y. Canby's fine article, "Are the Soviets Ahead in Space?" *National Geographic,* October 1986, 420–59.

5. "Russia's Space Program: Not Much More Than Cobwebs and Regrets," *New York Times,* March 21, 1995, p. C1.

6. M. Burgess and M. J. Perez, *Space Log: A Chronological Checklist of Manned Space Flights 1961–1990* (New York: Crown, 1991).

7. "Is NASA Necessary?" *Washington Post Magazine,* May 30, 1993, p. 10; also "Lost in Space: A \$1 Billion Satellite Is Gone, and So Is Another Chunk of NASA's Reputation," *Time,* September 6, 1993, p. 51.

8. Interview with Rainer Weiss.

9. Personal communication, January 7, 1996.

10. J. Houck, B. Soifer, M. Harwit, and J. L. Pipher, *Astrophysics Journal Letters* 178 (1972): 29.

11. Personal correspondence from Nancy Boggess, June 12, 1995.

12. Interviews with Nancy Boggess, Boulder, Colorado, August 31, 1994, and September 15, 1994.

13. Interview with Al Boggess, Boulder, Colorado, August 31, 1994.

CHAPTER 8: THE NEW AETHER DRIFT

1. Joseph Larmor, *Aether and Matter, Including a Discussion of the Influence of the Earth's Motion and Optical Phenomena* (Cambridge University Press, 1900).

2. Ronald W. Clark, *Einstein: The Life and Times* (World, 1971), p. 433.

3. D. W. Sciama, *Physical Review Letters* 18 (1967): 1065.

4. P. J. E. Peebles, *Physical Cosmology* (Princeton University Press, 1971).

5. P. J. E. Peebles, *Principles of Physical Cosmology* (Princeton University Press, 1993), p. 151.

6. Kurt Gödel, "An Example of a New Type of Cosmological Solution to Einstein's Field Equations of Gravitation," *Reviews of Modern Physics* 21 (1947): 447. In this article, Gödel also presents a theory of time travel based on the proposition that the universe is rotating. See also John Dawson, "Kurt Gödel in Sharper Focus," *The Mathematical Intelligencer* 6 (1984): 9.

7. E. K. Conklin (1961) and P. S. Henry (1971) detected tentative evidence of the dipole, which was later confirmed. See E. K. Conklin, *Nature* 222 (1969): 971; and P. S. Henry, *Nature* 231 (1971): 516.

CHAPTER 9: PARALLEL UNIVERSES

1. P. J. E. Peebles, "The Blackbody Radiation Content of the Universe and the Formation of Galaxies," *Astrophysical Journal* 142 (1965): 1317.

2. P. J. E. Peebles, "Primordial Helium Abundance and the Primordial Fireball II," *Astrophysical Journal* 146 (1966): 542–52.

3. J. Silk, "Fluctuations in the Primordial Fireball," *Nature* 215 (1967): 115.

4. J. Silk, "Cosmic Blackbody Radiation and Galaxy Formation," *Astrophysical Journal* 151 (1968): 459.

5. D. P. Woody, J. C. Mather, N. S. Nishioka, and P. L. Richards, "Measurement of the Spectrum of the Submillimeter Cosmic Background," *Physical Review Letters* 34 (1975): 1036–39.

6. Interview with Nancy Boggess, Boulder, Colorado, August 31, 1994.

CHAPTER 10: DUELING SATELLITES

1. A more comprehensive story of the IRAS project is Michael Rowan-Robinson, *Ripples in the Cosmos: A View Behind the Scenes of the New Cosmology* (W. H. Freeman/Spektrum, 1993).

2. Interviews with Nancy Boggess, Boulder, Colorado, August 31, 1994, and September 15, 1994.

3. R. O. Doyle, ed., "A Long-Range Program in Space Astronomy," NASA SP-213, July 1969.

4. "Outlook for Space," NASA SP-386, January 1976.

5. NASA publications expressed a more sympathetic view of the rewards of studying the cosmic background radiation than NASA officials during the 1960s and 1970s. See R. O. Doyle, ed., "A Long-Range Program in Space Astronomy,"

NASA SP-213 (1969); and "Outlook for Space," NASA SP-386 (1976). Both publications recommended that the cosmic background radiation be studied from a space platform.

6. P. J. E. Peebles, "The Black-Body Radiation Content of the Universe and the Formation of Galaxies," *Astrophysical Journal* 142 (1965): 1317; and "Microwave Radiation from the Big Bang," in *Relativity Theory and Astrophysics I: Relativity and Cosmology,* ed. J. Ehlers (American Mathematics Society, 1967).

7. S. Gulkis et al., *Cosmic Microwave Experiment Proposal,* Jet Propulsion Laboratory, unpublished, 1974.

8. Interview with Samuel Gulkis, Jet Propulsion Laboratory, Pasadena, California, February 15, 1995.

9. George Smoot and Keay Davidson, *Wrinkles in Time* (William Morrow, 1993), contains a lively discussion of Smoot's work with Luis Alvarez and their studies of the cosmic background radiation using high-altitude balloons and U-2 aircraft.

10. Interview with Phillip M. Lubin, Santa Barbara, California, February 13, 1995.

11. L. Alvarez et al., *Observational Cosmology: The Isotropy of the Primordial Black Body Radiation,* University of California, Berkeley, UCBSSL 556/75, unpublished, 1974.

12. Interview with Nancy Boggess, August 31, 1994.

CHAPTER 11: MATHEMATICAL MYTHOLOGY

1. John Mather, *Cosmic Background Explorer Satellite Newsletter #1,* NASA/Goddard Space Flight Center, 1976.

2. Steven Weinberg, *The First Three Minutes: A Modern View of the Origin of the Universe* (Basic Books, 1977), p. 72.

3. John Boslough, *Stephen Hawking's Universe* (William Morrow, 1985), pp. 52–53.

4. S. W. Hawking and R. Penrose, "The Singularities of Gravitational Collapse and Cosmology," *Proceedings of the Royal Society of London* A314 (1969): 529.

5. C. W. Misner, "The Mixmaster Universe," *Physical Review Letters* 22 (1969): 1071. A full explanation of the horizon problem is in Charles W. Misner, *Cosmology and Theology* (Freeman, 1977).

6. Interview with Alan Guth, Cambridge, England, 1982.

7. Alan Guth, "Inflationary Universe: A Possible Solution to the Horizon and Flatness Problems," *Physical Review D* 23 (1981): 347. For a good popular explanation of inflation, see Alan H. Guth and Paul J. Steinhard, "The Inflationary Universe," *Scientific American,* May 1984.

8. Interview with Stephen Hawking, Cambridge, England, 1983.

9. Stephen W. Hawking, "Is the End in Sight for Theoretical Physics?" an Inaugural Lecture published by Cambridge University Press, 1980 (pamphlet).

CHAPTER 12: DAY ONE

1. John Mather, Study Scientist, "Minutes of the COBE Team, First Meeting," Goddard Space Flight Center, June 28–29, 1976, unpublished.

2. Michael Janssen, "Interoffice Memorandum 325-MAJ-pm," Jet Propulsion Laboratory, August 16, 1976, unpublished.

3. John Mather, "Minutes of the COBE Team Meeting," Goddard Space Flight Center, August 8–9, 1976, unpublished.

4. Interview with Thomas L. Murdock, Danvers, Massachusetts, July 28, 1995.

5. R. G. Walker and S. D. Price, "AFCRL Infrared Sky Survey, Volume I. Catalog of Observations at 4, 11, and 20 Microns," Air Force Cambridge Research Laboratory Publication TR-75-0373, 1975.

6. M. G. Hauser, F. J. Low, G. H. Rieke, and S. D. Price, "The AFCRL Sky Survey: Supplementary Report to the Joint Scientific Mission Definition Team for an Infrared Astronomical Satellite (IRAS)," NASA Publication, June 1976.

7. Ibid., p. 22.

8. J. C. Mather, "Cosmic Background Explorer (COBE): Report to the Space Science Board," NASA/Goddard Space Flight Center, Greenbelt, Maryland, 1977.

9. Ibid., p. 3.

10. Interview with Edward L. Wright, University of California at Los Angeles, May 17, 1994.

11. Interview with Nancy Boggess, Boulder, Colorado, September 15, 1994.

12. Ibid.

CHAPTER 13: THE DIRECTORATE

1. Interview with Anthony Fragomeni, Timonium, Maryland, July 24, 1995.

2. Interview with Ray Weiss, Cambridge, Massachusetts, August 1, 1994.

3. Interview with Nancy Boggess, Boulder, Colorado, August 31, 1994.

4. Interview with Dennis K. McCarthy, Center for Astrophysical Sciences, Johns Hopkins University, Baltimore, Maryland, July 24, 1995.

5. Interview with Phil Lubin, Santa Barbara, California, February 13, 1975.

6. Interview with Dennis McCarthy.

CHAPTER 14: SKUNKWORKS

1. Interview with Nancy Boggess, Boulder, Colorado, September 15, 1994.

2. Interview with Dennis McCarthy, Baltimore, Maryland, July 24, 1995.

3. Engineering Directorate, Goddard Space Flight Center, "The Launch That Almost Never Was or COBE's Transformation from STS to Delta," *Engineering Newsletter* 8 (June 1990): 1.

4. Interview with Dennis McCarthy.

5. Ibid.

6. Interview with Olof Bengtson, Ball Aerospace, Boulder, Colorado, March 31, 1994.

7. R. A. Hopkins, "Thermal Performance of the COBE Superfluid Helium Dewar, As-built and with an Improved Support System," *Advances in Cryogenic Engineering* 33 (1988).

8. Tony Fragomeni and Dennis McCarthy, *COBE: A Case Study in Project Management,* Goddard Space Flight Center, July 1991.

9. "The Launch That Almost Never Was."

10. Interview with Anthony Fragomeni, Timonium, Maryland, July 24, 1995.

11. Interview with John Wolfgang, Goddard Space Flight Center, August 1, 1995.

CHAPTER 15: LOST IN SPACE

1. E. S. Cheng, P. R. Saulson, D. T. Wilkinson, and B. E. Corey, "Large-scale Anisotropy in the 2.7 K Radiation," *Astrophysics Journal Letters* 232 (September 15, 1979): L139−143.

2. S. P. Boughn, E. S. Cheng, and D. T. Wilkinson, "Dipole and Quadrupole Anisotropy of the 2.7 K Radiation," *Astrophysics Journal Letters* 242 (1981): L113−117.

3. See Alan Dressler, "The Large-Scale Streaming of Galaxies," *Scientific American* (1986): 27−38, for a general overview.

4. R. Scaramella et al., "A Marked Concentration of Galaxy Clusters," *Nature* 338 (1989): 562−64.

5. M. J. Geller and J. P. Huchra, "Mapping the Universe," *Science* 246 (1989): 897.

6. Interview with Edwin Turner, Princeton, New Jersey, 1992.

7. P. J. E. Peebles, *Science* 235 (1987): 372.

8. P. J. E. Peebles and Joseph Silk, "A Cosmic Book of Phenomena," *Nature* 346 (July 19, 1990): 233−39.

9. Interview with Richard Isaacman, Greenbelt, Maryland, September 16, 1991.

10. J. C. Mather et al., "A Preliminary Measurement of the Cosmic Microwave Background Spectrum by the *COBE* Satellite," *Astrophysical Journal* 354 (May 10, 1990): L37−40.

11. Warren E. Leary, "Spacecraft Sees Few Traces of a Tumultuous Creation," *New York Times,* January 14, 1990, p. A32.

12. Tim Folger, "Too Smooth a Universe," *Discover,* January 1991, pp. 34−35.

CHAPTER 16: THE PRIVATE LIFE OF THE COSMIC BACKGROUND RADIATION

1. Interview with Ned Wright, Los Angeles, May 17, 1994.

2. Interview with Charles Bennett, NASA Goddard Space Flight Center, Greenbelt, Maryland, August 16, 1995.

3. Ibid.

4. Interviews with Nancy Boggess, Boulder, Colorado, August 31, 1994, and September 15, 1994.

5. Interview with Mike Hauser, NASA Goddard Space Flight Center, Greenbelt, Maryland, July 26, 1995.

6. Interviews with Nancy Boggess and Chuck Bennett.

7. Jeffrey Kahn, "Seeds of the Modern Universe Detected in the Ancient Microwave Sky," Lawrence Berkeley Laboratory, Public Information Department, Berkeley, California, 1992.

8. Personal correspondence from Ray Weiss, January 6, 1996.

9. The four papers the COBE team published relating to the DMR results were: G. F. Smoot et al., "Structure in the COBE Differential Microwave Radiometer First-year Maps"; C. L. Bennett et al., "Preliminary Separation of Galactic and Cosmic Microwave Emission for the COBE Differential Microwave Radiometer"; E. L. Wright et al., "Interpretation of the Cosmic Microwave Background Radiation Anisotropy Detected by the COBE Differential Microwave Radiometer" and A. Kogut et al., "COBE DMR: Preliminary Systematic Error Analysis" —all in *Astrophysical Journal* 396 (September 1, 1992).

10. George Smoot and Keay Davidson, *Wrinkles in Time* (Morrow, 1993), p. 288.

11. Ibid., p. 289.

12. *Wall Street Journal,* April 24, 1992.

13. John Horgan, "COBE's Cosmic Cartographer," *Scientific American,* July 1992, p. 41.

14. "COBE Causes Big Bang in Cosmology," *Science News,* May 6, 1992, p. 292.

15. "Big Bang Echoes Through the Universe," *New Scientist,* May 2, 1992, p. 4.

16. Horgan, "COBE's Cosmic Cartographer," p. 41.

17. Michael White, "Eureka! They Like Science," *The Sunday* [London] *Times,* December 13, 1992.

18. Interview with George Smoot, Lawrence Berkeley Laboratory, Berkeley, California, May 19, 1995.

19. Interview with Ned Wright.

20. Interview with Sam Gulkis, Pasadena, California, February 15, 1994.

21. Interview with Nancy Boggess, August 31, 1994.

22. Personal correspondence from Ray Weiss, January 6, 1996.

23. Interview with Charles Bennett.

24. George Smoot, Letter to members of the COBE science working group, September 1, 1992.

Chapter 17: The Cosmos After COBE

1. Floyd Stecker, "Ultrahigh Energy Photons, Electrons, and Neutrinos: The Microwave Background and the Universal Cosmic-Ray Hypothesis," *Astrophysics and Space Science* 20 (January 5, 1973): 47–57.

2. Interview with P. J. E. Peebles, Princeton, New Jersey, December 20, 1994.

3. E. L. Wright et al., "Interpretation of the COBE FIRAS CMBR Spectrum"; J. C. Mather et al., "Measurement of the Cosmic Microwave background Spectrum by the COBE FIRAS Instrument"; D. J. Fixsen et al., "Cosmic Microwave Background Dipole Spectrum Measured by the COBE FIRAS Instrument"; and D. J. Fixson et al., "Calibration of the Far Infrared Absolute Spectrophotometer (FIRAS) on the Cosmic Background Explorer (COBE)" all in *Astrophysical Journal* 420 (January 10, 1994).

4. John Noble Wilford, "Scientists Say Satellite's Findings Confirm Big Bang Cosmic Theory," *New York Times,* January 8, 1993, p. A1.

5. D. J. Fixsen et al., "The Cosmic Microwave Background Radiation Spectrum from the Full COBE FIRAS Data Set," *Astrophysical Journal,* 1996, accepted for publication.

EPILOGUE

1. See John D. Barrow and Frank J. Tipler, *The Anthropic Cosmological Principle* (Oxford University Press, 1988).

2. George W. Collins, "Quo Vadis, Astronomia?" *Astrophysical Journal* 449 (August 20, 1995): 399–403.

A BRIEF COBE GLOSSARY

AAAS   American Association for the Advancement of Science, the largest organization of professional scientists in the United States, covering all areas.

AAS   American Astronomical Society.

absolute zero   The temperature, -273° Celsius, at which all molecular activity halts.

anisotropy   Differences from one direction to another, from the Greek *an-iso,* meaning "not the same."

APS   American Physical Society.

blackbody   An object that absorbs all radiation that falls on it. If it is all at one single temperature, it also emits radiation according to Planck's simple formula. Its brightness is dependent only on wavelength and temperature.

bolometer   In infrared astronomy, a detector that measures radiation at all wavelengths by converting it to heat and measuring the temperature change.

calibrator   A device that produces a known amount of input to a measuring instrument so that measured data can be correctly interpreted and converted to meaningful units.

COBE   Cosmic Background Explorer, an orbiting satellite with three scientific instruments on board.

cryostat and/or dewar   A large vacuum-insulated flask, like a Thermos bottle, used for storing cold liquids (cryogens).

dipole   Something with two ends (poles); in cosmology, a pattern on the sky that changes gradually from +1 in one direction to -1 in the opposite direction.

DIRBE   Diffuse Infrared Background Experiment. One of the three COBE instruments, designed to measure the infrared radiation from the earliest objects to form after the Big Bang as well as infrared radiation from everything else, including stars and dust in the Milky Way, other galaxies, and interplanetary dust. It works at 10 wavelengths from 1.2 to 240 micrometers, and is cooled to 1.5° Kelvin by liquid helium.

DMR   Differential Microwave Radiometers. One of the three COBE instruments, designed to measure the anisotropy of the cosmic microwave background radiation; it works at 3 wavelengths, 3.3, 5.7, and 9.6 mm. The 3.3 and 5.7 mm receivers operate at 140° K, and the 9.6 mm receivers at room temperature.

Explorer A series of relatively small NASA research satellites, managed by NASA within a budget allocation set by Congress.

FIRAS Far Infrared Absolute Spectrophotometer. One of the three COBE instruments, designed to measure the spectrum of the cosmic background radiation and compare it with the predicted blackbody spectrum; it is cooled to $1.5°$ K by liquid helium, and measures wavelengths from 0.1 to 10 mm. It uses an interferometer to determine wavelengths.

inflation An idea to explain the initial phase of the expanding universe. Certain subatomic particles might exist that could cause extremely rapid exponential expansion of the early material. Inflation would explain the general uniformity of the present universe.

interferometer A device that divides incoming waves into two (or more) parts and then combines them again. When they combine, the wave parts interfere with each other, to give intensity patterns that depend on the wavelength. This effect can be used to measure the wavelengths.

IRAS Infrared Astronomical Satellite. A project of the United States, the Netherlands, and the United Kingdom to map the infrared sky. It was launched in 1983 and used a liquid helium tank and detectors similar to those on COBE.

liquid helium The coldest known liquid, it can be cooled to temperatures around 1 to $4°$ K. Helium is an inert gas found in the sun and as an impurity in natural gas from certain places. Its national supply is endangered because Congress is considering selling off the gas wells from which it comes.

matrix management A method of managing a large organization in which specialists are assigned to groups divided according to kinds of skills, that is, mechanical engineers in one group, electronic engineers in another, technicians in another, and so forth. When a project is undertaken, the managers of these groups of specialists assign the work to individuals in their groups. It is the opposite of the integrated project team, in which individuals are assigned to a project and take all their direction from the project management.

photon A particle of light. Light has both wave and particle properties. It can be divided and suffer interference like a wave, but in a particle detector it can release individual electrons as though it were a stream of particles.

PI Principal Investigator. The person responsible for a scientific investigation; this individual decides how to spend resources and what steps should be taken.

quantum mechanics   A branch of physics dealing with the wave nature of particles. The waves are described by complex numbers, and the squares of these complex numbers are interpreted as probabilities. The equations are simple and agree with measurements, but the interpretation is hotly debated by philosophers and physicists. Einstein was bitterly opposed to the probability idea of quantum mechanics and famously stated, "God does not play dice with the universe."

redshift   In cosmology, a change of the measured wavelength of light due to the expansion of the universe and the speed of recession of the source of light.

BIBLIOGRAPHY

As good almost kill a man as kill a good book.

—JOHN MILTON, *Areopagitica,* 1644

Books and articles likely to be of interest to the general reader are included here, while the endnotes contain technical books and scientific papers.

A series of six news articles in *Science* magazine, vol. 272, pp. 1426–37, 1996, summarizes the state of affairs of modern cosmology very well.

Armitage, Angus. *Copernicus: The Founder of Modern Astronomy.* New York: Barnes, 1962.

Barrow, John D. *The Origin of the Universe.* New York: Basic Books, 1994.

Barrow, John D., and Frank J. Tipler. *The Anthropic Cosmological Principle.* Oxford: Oxford University Press, 1988.

Barrow, John D., and Joseph Silk. *The Left Hand of Creation: The Origin and Evolution of the Expanding Universe.* New York: Basic Books, 1983.

Berger, A., ed. *The Big Bang and Georges Lemaître.* Amsterdam: Reidel Publishing Co., 1984.

Bernstein, Jeremy. *Einstein.* New York: Viking, 1973.

Bernstein, Jeremy, and Gerald Feinberg, eds. *Cosmological Constants: Papers in Modern Cosmology.* New York: Columbia University Press, 1988. From before Einstein to cosmic strings.

Bondi, Hermann. *Assumption and Myth in Physical Theory.* Cambridge: Cambridge University Press, 1967.

Boslough, John. *Stephen Hawking's Universe.* New York: William Morrow, 1985. A brief history of the life of the famed cosmologist.

Broad, William, and Nicholas Wade. *Betrayers of the Truth: Fraud and Deceit in the Halls of Science.* New York: Simon & Schuster, 1982.

Bronowsky, Jacob. *The Ascent of Man.* Boston: Little, Brown, 1973. A classic study of humanity's self-discovery.

Butterfield, Herbert. *The Origins of Modern Science, 1300–1800.* New York: Macmillan, 1951. A classic work.

Calder, Nigel. *Einstein's Universe.* New York: Greenwich House, 1979. A crystal-clear exposition of relativity for the lay reader.

Canby, Thomas Y. "Are the Soviets Ahead in Space?" *National Geographic,* October 1986.

Chaikin, Andrew. *A Man on the Moon.* New York: Viking, 1994.

Chapman, Clark R., and David Morrison. "Impacts on the Earth by Asteroids and Comets: Assessing the Hazard." *Nature* 367 (1994).

Chown, Marcus. *Afterglow of Creation: From the Fireball to the Discovery of Cosmic Ripples.* London: Arrow Books, 1993, 1996.

Cornell, James, ed. *Bubbles, Voids, and Bumps in Time: The New Cosmology.* Cambridge: Cambridge University Press, 1988. A collection of popular articles by cosmologists.

Crawford, I. A. "Space, World Government and 'The End of History.'" *Journal of the British Interplanetary Society* 46 (1993).

Davies, Paul. *God and the New Physics.* New York: Simon & Schuster, 1983.

Eddington, A. S. *The Nature of the Physical World.* New York: Macmillan, 1928. How the universe was perceived on the eve of Hubble's discovery of galactic recession.

Einstein, Albert. *The Meaning of Relativity.* Princeton: Princeton University Press, 1955. One of Einstein's most accessible works on the subject, based on his 1922 lectures.

Ferris, Timothy. *Coming of Age in the Milky Way.* New York: William Morrow, 1988. A literary account of humanity's struggle to understand the universe.

Feynman, Richard P. *Surely, You're Joking, Mr. Feynman!* New York: Norton, 1985. Delightful reading from the great physicist.

Gamow, George. *One, Two, Three . . . Infinity.* New York: Viking, 1947. The book that inspired thousands of students in the days when the Big Bang was new.

———. *The Creation of the Universe.* New York: Viking, 1957. Of historical interest, mostly.

Geymonat, Ludovico. *Galileo Galilei: A Biography and Inquiry into His Philosophy of Science.* Trans. Stillman Drake. New York: McGraw-Hill, 1965. A classic study.

Gillispie, Charles Coulston. *The Edge of Objectivity: An Essay in the History of Scientific Ideas.* Princeton: Princeton University Press, 1960. A lively and intellectual account.

Gott, J. Richard III. "Implications of the Copernican Principle for Our Future Prospects." *Nature* 263 (1993).

Gribbin, John. *In the Beginning: After COBE and Before the Big Bang.* Boston: Little, Brown, 1993.

Gulkis, Samuel, Phillip M. Lubin, Stephan S. Meyer, and Robert F. Silverberg. "The Cosmic Background Explorer." *Scientific American,* January 1990. A good general description of the instruments on board from four who should know.

Guth, Alan H., and Paul J. Steinhardt. "The Inflationary Universe." *Scientific American,* May 1984. Probably the best short explanation of the subject for the general reader.

Harwit, Martin. *Cosmic Discovery: The Search, Scope, and Heritage of Astronomy.* New York: Basic Books, 1981. How astronomy is done, quantified.

Hawking, Stephen W. *A Brief History of Time.* New York: Bantam, 1988.

Hoyle, Fred. *The Nature of the Universe.* New York: Harper, 1950. The cosmos without the Big Bang.

Krauss, Lawrence M. "Dark Matter in the Universe." *Scientific American,* December 1986. A condensed explanation.

———. *The Fifth Essence: The Search for Dark Matter in the Universe.* New York: Basic Books, 1989.

Kuhn, Thomas S. *The Copernican Revolution.* Cambridge: Harvard University Press, 1970.

———. *The Structure of Scientific Revolutions.* Chicago: University of Chicago Press, 1970. Second edition of the work that redefined the history of science and added *paradigm* to the general lexicon.

Lemaître, G. *The Primeval Atom.* New York: Van Nostrand, 1950. One of the earliest expositions on the Big Bang.

Lerner, Eric J. *The Big Bang Never Happened: A Startling Refutation of the Dominant Theory of the Origin of the Universe.* New York: Times Books, 1991. Promotion of an alternative theory.

Leslie, John, ed. *Physical Cosmology and Philosophy.* New York: Macmillan, 1990. Musings from George Gamow and others.

Lewis, John S., and Ruth A. Lewis. *Space Resources: Breaking the Bonds of Earth.* New York: Columbia University Press, 1994.

Lightman, Alan, and Roberta Brawer, eds. *Origins: The Lives and Worlds of Modern Cosmologists.* Cambridge: Harvard University Press, 1990. Often fascinating interviews with many of cosmology's leading lights.

Linde, Andrei. "The Self-Reproducing Inflationary Universe." *Scientific American,* November 1994, pp. 45–48. A lovely explanation of some radical but plausible mathematical explorations of the universe before the Bang.

McConnell, Malcolm. *Challenger: A Major Malfunction.* New York: Doubleday, 1990. Political intrigues that may have led to the shuttle's explosion.

McDougall, Walter A. *The Heavens and the Earth: A Political History of the Space Age.* New York: Basic Books, 1985.

North, John. *Isaac Newton.* Oxford: Oxford University Press, 1967. A good explanation of Newton's mathematics for the general reader.

———. *The Norton History of Astronomy and Cosmology.* New York: Norton, 1994. Contains a brief account of COBE and its significance.

Overbye, Dennis. *Lonely Hearts of the Cosmos.* New York: HarperCollins, 1991. An entertaining account of modern cosmology and cosmologists.

Partridge, R. Bruce. *3K: The Cosmic Microwave Background Radiation.* New York: Cambridge University Press, 1994. A scholarly tour de force that is enjoyable to read. Partridge joined Dicke's Gravity Group in 1965. He summarizes five decades of work on the CMBR.

Peebles, P. J. E. *Physical Cosmology.* Princeton: Princeton University Press, 1970. Contains an account of the discovery of the cosmic background radiation.

———. *Principles of Physical Cosmology.* Princeton: Princeton University Press, 1993. The standard textbook.

Peratt, Anthony L. "Not with a Bang: The Universe May Have Evolved from a Vast Sea of Plasma." *The Sciences,* January–February, 1990. A good nontechnical article detailing the plasma universe of Hannes Alfvén.

Popper, Karl, *The Logic of Scientific Discovery.* New York: Harper & Row, 1968. The classic work from the renowned philosopher of science.

Rowan-Robinson, Michael. *Ripples in the Cosmos: A View Behind the Scenes of the New Cosmology.* New York: W. H. Freeman, 1993. An account of the IRAS mission.

Rubin, Vera C. "Dark Matter in Spiral Galaxies." *Scientific American,* December 1988. A good general description of its discovery and implications.

Sagan, Carl. *Pale Blue Dot: A Vision of the Human Future in Space.* New York: Random House, 1994. The future in the inimitable Sagan style.

Silk, Joseph. *The Big Bang.* San Francisco: Freeman, 1989. Discusses the big issues in cosmology today.

Smoot, George, and Keay Davidson. *Wrinkles in Time.* New York: William Morrow. Contains an account of the COBE discoveries.

Tsiolkovsky, Konstantin. *The Call of the Cosmos.* Moscow: Foreign Languages Publishing House, 1960. Trans. into English. The story of the early space pioneer.

Turner, Edwin L. "Gravitational Lenses." *Scientific American,* July 1988. An excellent nontechnical overview.

Weinberg, Steven. *The First Three Minutes: A Modern View of the Origin of the Universe.* New York: Basic Books, 1977. Still an excellent introduction to the early universe.

Wheeler, John Archibald. *A Journey into Gravity and Spacetime.* New York: Scientific American Library, 1990. A wonderful excursion with the master teacher emeritus.

White, Andrew Dickson. *A History of the Warfare of Science with Theology in Christendom.* New York: Macmillan, 1965. The interminable battle rages on, with science apparently beginning to take the upper hand.

Wilford, John Noble. *Mars Beckons: The Mysteries, the Challenges, the Expectations of Our Next Great Adventure in Space.* New York: Knopf, 1990.

WHERE ARE THEY NOW?

CHUCK BENNETT is a branch head and is PI for the MAP, the Microwave Anisotropy Probe selected as a successor to the COBE DMR.

NANCY BOGGESS retired from NASA and moved to Boulder, Colorado, with her husband, Al; they travel frequently to distant lands in pursuit of exotic species of birds.

TONY FRAGOMENI retired from NASA and lives in Timonium, Maryland; he maintains his sense of humor and stays in touch with his former colleagues, especially Dennis McCarthy.

MIKE HAUSER was named deputy director of the Space Telescope Science Institute in Baltimore; he has continued working on DIRBE science and data processing effort.

JOHN MATHER remains at Goddard, where he still analyzes COBE data; he has been named study scientist of the successor to the Hubble Space Telescope.

DENNIS MCCARTHY left Goddard and began working at Johns Hopkins University in Baltimore, where he still develops new spacecraft for NASA.

GEORGE SMOOT was named a professor at the University of California at Berkeley, and received the E.O. Lawrence award from the Lawrence Berkeley Laboratory in 1995; he has

been involved in a new balloon experiment.

RAY WEISS has moved on to his other love: gravity wave antennas. He had a minor heart attack in 1995, which slowed him down from working sixteen hours a day, seven days a week, for months at a time.

DAVE WILKINSON has remained on the physics faculty at Princeton University and continued to lead experiments to measure the cosmic background radiation. He leads the MAP effort with Chuck Bennett.

JOHN WOLFGANG has continued working for the Engineering Directorate at Goddard, teaching young engineers how to build spacecraft to NASA specifications.

NED WRIGHT has been a full professor at the University of California at Los Angeles since 1984; he also serves as an associate editor of the *Astrophysical Journal.*

MANY OF COBE'S FINEST ENGINEERS have risen to high positions at Goddard and elsewhere throughout NASA; some retired and have gone into the commercial sector.

THE COBE spacecraft remains in orbit. As far as is known it is still working, but its job is done.

INDEX

DISCARDED